Cover: The busy little mining camp of Sherman was located in a lush wooded valley below Sunshine Peak, approximately sixteen miles southwest of Lake City. Due to its location nearly midway between Lake City and Animas Forks, Sherman was a convenient stagecoach stop. Behind the store, the Black Wonder Mill can be seen in the background. *Collection of Dave Southworth.*

Colorado Mining Camps

by Dave Southworth

Wild Horse Publishing

> *With heartfelt appreciation, I dedicate this book to my mother, Ruth S. Bennett, for her many hours of selfless devotion to this project.*
>
> **Dave Southworth**

Cover design by Dave Southworth
Maps by Dave Southworth & United States Geological Survey

Library of Congress Cataloging-in-Publication Data

Southworth, Dave, 1937–
  Colorado Mining Camps

  Bibliography: p. 303
  Includes index.
  1. Colorado—History 2. Mining Camps—Colorado—History.
  3. Frontier and pioneer life—Colorado—History. 4. Cities and towns—Colorado—History.
  5. Mines and mineral resources—Colorado—History. 6. Colorado—Description and travel—Guide-books. I. Southworth, Dave II. Title.
  ISBN: 978-1-890778-00-2
  ISBN: 1890778001
Copyright © 2011 by Wild Horse Publishing.
Printed in the United States of America.

All rights reserved. No part of this book may be reproduced, stored in a retrieval system or transmitted in any form or by any means electronic, mechanical, photocopying, recording or otherwise, without the prior permission of Wild Horse Publishing.

Contents

Preface 9
Brief History 11
Map: State of Colorado 13

NORTH CENTRAL REGION 14

GILPIN COUNTY 15
    Gilpin County Map, Mountain City, Central City, Black Hawk,
    Nevadaville, Russell Gulch, Perigo, Rollinsville, Wide Awake,
    Baltimore, Apex, American City, Gilpin

CLEAR CREEK COUNTY 30
    Clear Creek County Map, Idaho Springs, Dumont, Empire,
    North Empire, Waldorf, Georgetown, Silver Plume, Lawson,
    Silver Creek, Freeland, Alice, Yankee, Ninety Four. Lamartine

BOULDER COUNTY 52
    Boulder County Map, Gold Hill, Jamestown, Ward, Caribou,
    Cardinal, Nederland, Salina, Sunshine, Crisman, Sunset,
    Wallstreet, Magnolia, Eldora, Hessie, Summerville, Camp Tolcott

LARIMER COUNTY 70
    Larimer County Map, Manhattan

DOUGLAS COUNTY 72
    Douglas County Map, West Creek

CENTRAL REGION 75

PARK COUNTY 76
    Park County Map, Fairplay, Buckskin Joe, Montgomery,
    Park City, Tarryall, Hamilton, London Junction, Alma,
    Jefferson, Como, Horseshoe, Leavick, Balfour, Guffey,
    Webster

LAKE COUNTY 94
    Lake County Map, Leadville Area Map, Oro City, Leadville,
    Stringtown, Adelaide, Stumptown, Twin Lakes, Malta,
    St. Kevins, Tabor City, Everett, Brumley

TELLER COUNTY 109
    Teller County Map, Cripple Creek Area Map, Cripple Creek,
    Victor, Goldfield, Gillett, Independence, Elkton, Cameron,
    Altman, Stratton, Anaconda

SOUTH CENTRAL REGION 127

CHAFFEE COUNTY 128
    Chaffee County Map, Cache Creek, Granite, Hortense,
    Harvard City, Babcock, Monarch, Arbourville, Shavano,
    Alpine, Romley, Hancock, Stonewall, Saint Elmo,
    Buena Vista, Maysville, Iron City, Beaver City, Vicksburg,
    Rockdale, Winfield, Turret

FREMONT COUNTY 146
    Fremont County Map, Whitehorn

CUSTER COUNTY 148
    Custer County Map, Rosita, Querida, Ilse, Silver Cliff,
    Westcliffe

SAGUACHE COUNTY 155
    Saguache County Map, Bonanza, Liberty, Crestone, Iris

RIO GRANDE COUNTY 160
    Rio Grande County Map, Summitville, Jasper

CONEJOS COUNTY 163
    Conejos County Map, Platoro

COSTILLA COUNTY 166
    Costilla County Map, Russell

MINERAL COUNTY 168
    Mineral County Map, Bachelor, North Creede,
    Creede, Spar City

NORTHWEST REGION 175

SUMMIT COUNTY 176
    Summit County Map, Breckenridge, Parkville, Delaware Flats,
    Lincoln, Preston, Tiger, Saints John, Montezuma, Argentine,
    Chihuahua, Keystone, Masontown, Frisco, Conger Camp,
    Boreas, Dyersville, Robinson, Kokomo, Rexford, Swandyke

PITKIN COUNTY 198
   Pitkin County Map, Independence, Ashcroft, Aspen,
   Highland, Lenado, Emma, Ruby, Redstone

EAGLE COUNTY 209
   Eagle County Map, Redcliff, Gold Park, Holy Cross City,
   Basalt, Gilman, Fulford

ROUTT COUNTY 217
   Routt County Map, Hahns Peak, Columbine

JACKSON COUNTY 220
   Jackson County Map, Teller City, Pearl

GRAND COUNTY 223
   Grand County Map, Lulu, Arrow

SOUTHWEST REGION 226

SAN JUAN COUNTY 227
   San Juan County Map, Silverton, Mineral Point,
   Howardsville, Eureka, Animas Forks, Gladstone

OURAY COUNTY 238
   Ouray County Map, Ouray, Sneffels, Red Mountain, Ironton,
   Guston, Camp Bird

GUNNISON COUNTY 249
   Gunnison County Map, Tin Cup, Pitkin, Ohio City, White Pine,
   North Star, Tomichi, Alpine Station, Hillerton, Crested Butte,
   Irwin, Gothic, Schofield, Crystal, Marble, Chance, Vulcan,
   Baldwin, Dorchester, Bowerman

HINSDALE COUNTY 270
   Hinsdale County Map, Lake City, Henson, Capitol City,
   Sherman, Whitecross, Carson

SAN MIGUEL COUNTY 278
   San Miguel County Map, Telluride, Placerville, Ophir,
   Alta, Ames, Pandora, Tomboy

DOLORES COUNTY 289
   Dolores County Map, Rico, Dunton

LA PLATA COUNTY 293
    La Plata County Map, Parrott City, La Plata

Glossary of Mining Terms 295
Acknowledgements 299
Bibliography 303
Index 313

Other Works by Dave Southworth

──────────── Books ────────────

Feuds on the Western Frontier
Gunfighters of the Old West
Colorado Gold Dust: Short Stories and Profiles
Ghost Towns and Mining Camps of the San Juans

──────────── Videos ────────────

*Colorado Mining Camps: A Pictorial Treasure of the
    Gold and Silver Boom*
Leadville: The Boom Years
Mining Camps of the San Juans
Cripple Creek and the Mining Camps of Teller County
The Mining Camps of Northwest Colorado
The Mining Camps of Gilpin and Clear Creek Counties
The Mining Camps of South Central Colorado
Boulder County Mining Camps: A Look Back

Dave Southworth

Preface

Spread across Colorado's high country are forgotten cities and others which will never be forgotten — many of which are stocked with reminders of the fabulous gold rush days. Part of our heritage lies with the hardy people of a century ago and the towns and mines where they lived and worked. This book presents over 200 towns which were a part of the turbulent gold and silver era. There were colorful characters of all kinds — heroes and heroines, con artists and gamblers, itinerate preachers and parlor girls, tin stars and gunslingers, thieves and fools, braggarts and optimists. This is their story — and these are their towns.

Each year thousands of travelers journey to see what still exists from some of the most interesting pages in Colorado history. The mining camps and mines are attractions to nostalgists, adventurers, historians, tourists, or anybody who loves interesting things and places. My objective in writing this book was to create a comprehensive history of (and guide to) those mining camps which were a part of the Colorado mining boom. Excluded are towns with insignificant history, whose names are simply epitaphs for places that died and disappeared. Many of the cities and towns which played an important part in the mining boom have been inhabited continually since they were established. Conversely, there are some in various stages of collapse, and many which have totally disappeared. Some became ghost towns but withstood the test of time and have had a resurgence. Each community has its own special story, and they are often spiced with legend, lore and humor.

There is some truth in the expression "a picture is worth a thousand words." Naturally it depends on the picture. Sometimes it is easiest to depict life in a roaring mining town with an old photograph. When possible, photos were selected which included people. Folks usually dressed up in their Sunday best when a photographer was scheduled to be in town.

Many maps are included as a reference for those who wish to venture into the high country. Each site may be located by using the geographical description (and directions) in conjunction with the proper map. U.S. Forest Service maps offer more detail of any area. When seeking a remote site, it might be helpful to travel with a topographic map and compass.

Many mining camps are accessible by automobile. Others may be reached with a 4-wheel drive vehicle. An experienced mountain trail horse, trail bike, or a back-pack hike is recommended travel to a couple of locations. Sometimes it is necessary to cross a labyrinth of jeep trails and pack trails to reach certain destinations. In cases like this it wise to get very specific directions or take a guide. Guides, outfitters, and jeep tours are generally expensive. Most are reputable and interesting. Caution is advised before making any trip, however, because some tours will literally "take you for a ride." Find out in advance exactly where the tour goes, what will actually be seen, how long it will take, and the cost.

Traveling into the high country can be great fun, but it is not for everybody. If such a trek is a hardship, much may not have been missed for some of these remote sites are in locations where heavy snows over the years have barely left traces of once active mining camps. This book should be valuable for those "armchair explorers" who will never physically visit these sites, but who will be able through this text to slip back in their imaginations to this colorful period of history.

Mines, mining towns, and spectacular scenery certainly go together. Colorado is rich in minerals and rich in pastoral splendor. Nature's serenity seems to stretch on forever across the awesome mountain peaks and down through luscious valleys. Most of the mining towns are (or were) located in places of great beauty. To tour them is exhilarating. The flapping shutters and creaking boards will amplify your imagination — so should this book!

Dave Southworth

Brief History

In 1806, the United States government dispatched a military man, Zebulon Montgomery Pike, to explore part of what is now Colorado in an effort to find the river boundary between the United States and Spanish territory. West of the Mississippi River lay wide open unexplored country. Spain claimed a large part of this territory, but there was uncertainty over the exact location of the boundary lines. During his expedition, Pike discovered the beginnings of the South Platte and Arkansas Rivers, and a huge mountain on the horizon. The explorers traveled for days in an effort to reach the mountain, but it seemed to get no closer. As snow and freezing temperatures set in, Pike had to abandon his trek. Pike never climbed the mountain, nor did he name it for himself. Trappers and hunters dubbed it "Pike's Peak," and the name stuck.

During the summer of 1850, Lewis Ralston and a band of southern prospectors discovered gold at Ralston Creek near its confluence with Clear Creek. Although this was the first known find in Colorado, it didn't amount to much. Ralston and his group pushed on toward the gold fields of northern California.

Ten years after the California Gold Rush, William Green Russell and his party from Georgia discovered gold at Dry Creek in 1858. Hundreds of prospectors, with picks and shovels, flocked to the camps of Denver City and Auraria. The cry was "Pike's Peak or Bust." They broke rocks and ravaged streambeds with little to show for it. Most left in disgust calling the strike a hoax. They were hasty in their decision, however, for gold strikes by George Jackson near Idaho Springs and John Gregory near Central City in the spring of 1859 brought thousands more stampeding into the territory. By summer, over 5,000 men were working the Gregory Lode with more than 100 sluices. Miners were inspired to spread out seeking gold deposits in other parts of the territory. As discoveries were made, towns cropped up throughout the mountains. Denver quickly grew as a supply town for the mining camps. More immigrants headed west taking with them their hopes, memories, and whatever they could stuff into their wagons.

In February 1861, the Colorado Territory was established. The Anglo population of Colorado was set at 25,242 when Governor William Gilpin took the first head count that same year. Men outnumbered women by a margin of thirty to one.

Mining towns popped up as tent cities. Gradually the tents were replaced with log cabins constructed of squared-hewn timber. When a saw mill was completed, frame structures were built, many with tall and massive false fronts. Stores were erected shoulder-to-shoulder so the imposing fronts could hide the buildings behind them. As a town showed signs of permanence, some stone and brick buildings were constructed. Many finer homes were built of Victorian architecture. In certain areas, many were trimmed with elaborate gingerbread.

The earliest towns had sanitation problems. Much trash was thrown into the streets. Mules, horses, and cattle contributed to the refuse. Personal cleanliness was minimal. Miners worked up a sweat during the daytime, then slept in their clothes, for tent walls were thin and mountain nights were chilly. Many never bothered to bathe, especially during colder weather. As a result, epidemics were common. As families arrived in later migrations, most towns became clean and comfortable villages.

Placer mining was most common in the early 1860s. Prospectors scrambled up and down the streams in search of color. When successful, they staked their claims and panned the stream beds. Others combed the mountains in search of exposed ore or float gold. It wasn't long before all of the easily obtainable gold had been found, so miners began tunneling into the mountain sides, sinking shafts to reach their veins. Gold found in this manner was difficult to extract from the ore. Prof. Nathaniel Hill solved the ore reduction problem when he constructed his Boston and Colorado Smelter at Black Hawk in 1868.

Prospectors pushed deeper into the mountains discovering new mining areas. During the seventies and eighties, mining really boomed. When silver was discovered in Leadville in 1877 — the year after Colorado achieved statehood — the population of Lake County grew from a handful to nearly twenty-four thousand. In a few short months, silver became more dominant than gold throughout the state.

For years the Ute Indians who roamed the mountains of Colorado remained friendly with white settlers. Their patience was stretched, however, as more prospectors moved into the mountains in search of new mining areas. The United States government made several agreements with Chief Ouray which resulted in a reduction of their territory. During an uprising at the White River Ute Indian Agency in 1879, a band of Utes killed Nathan Meeker, an Indian agent, in what became known as the Meeker Massacre. As a result, the tribe was relocated further west, predominately into Utah, virtually giving white settlers a reign over all of Colorado.

The population explosion in the mining towns was abetted by a maze of railroads which snaked through valleys and over mountains bringing supplies and more people. The railroads also provided a means of shipping ore from the mines to the smelters with a lot less hardship.

As mining moved deeper beneath the surface of the earth, more capital was required. Mining companies, capitalized in the east or even in Europe, pumped money into mining operations. Corporate organization of lucrative claims occurred. In many instances, the original claim holder became an employee of the corporation. Hydraulic mining became popular in certain areas.

Repeal of the Sherman Silver Purchase Act in 1893 created a devaluation of silver and dealt a devastating blow to the mining industry. Many mines were shutdown, communities became ghost towns, and railroads were abandoned. Although the bonanza had not yet occurred in the Cripple Creek-Victor area, gold production throughout much of the state was declining by this time also. The remains of mines, old buildings, and ghost towns bring back the days of yesteryear — a colorful page in Colorado history.

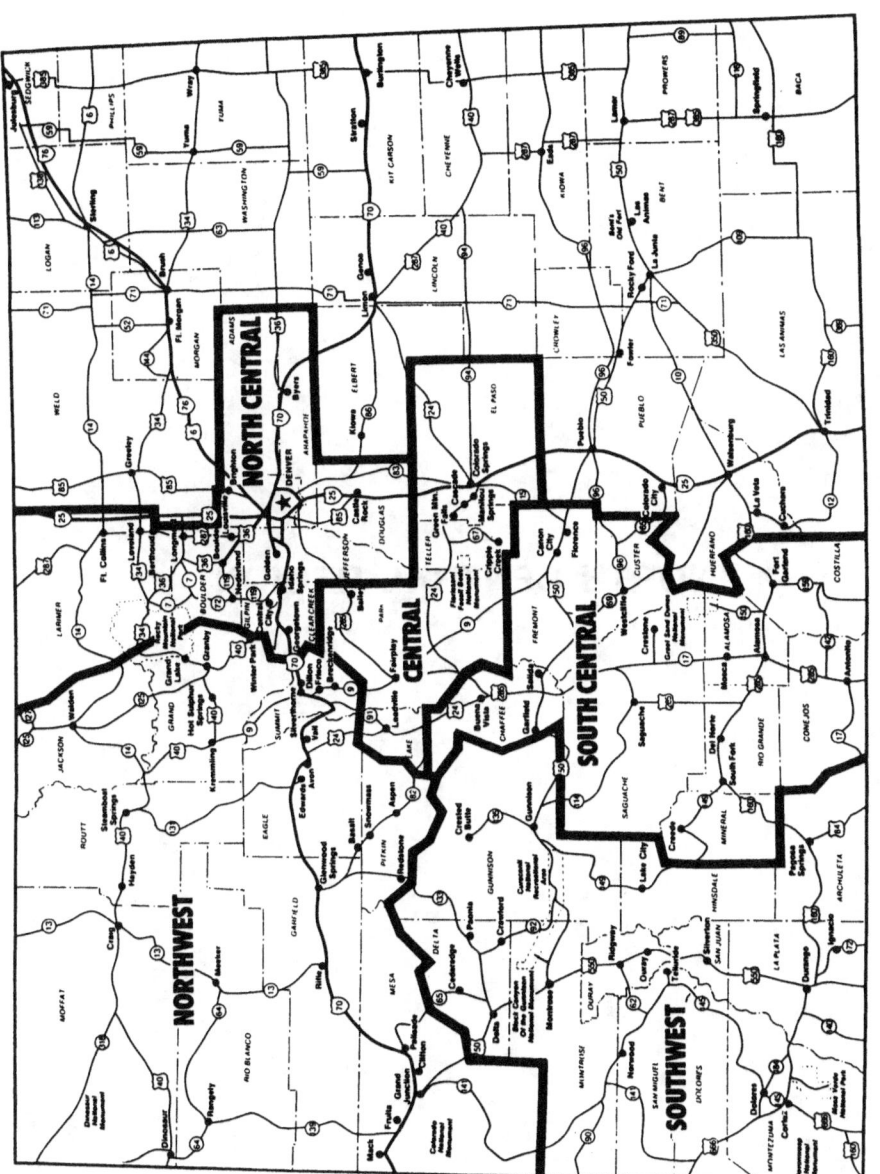

STATE OF COLORADO

NORTH CENTRAL REGION

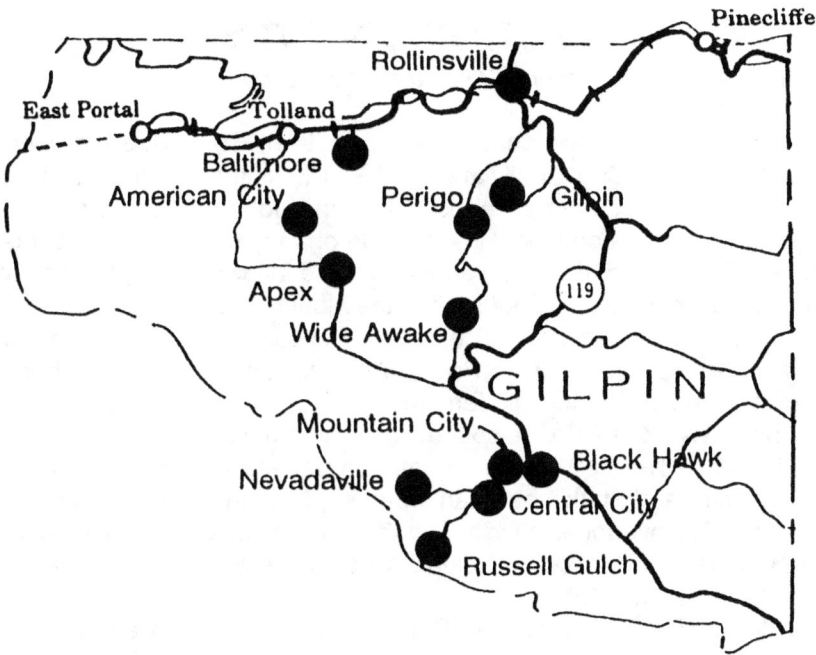

GILPIN COUNTY

"Law is five cartridges in the cylinder —justice is the one in the chamber."

MOUNTAIN CITY
Location: 1/2 mile east of Central City

Mountain City was founded near the site of Colorado's first lode gold discovery. Between Central City and Black Hawk (neither of which existed at the time) stands a monument which is inscribed, "John H. Gregory of Georgia Discovered the First Lode Gold in Colorado on May 6, 1859." That same year Mountain City sprang to life. By January of the following year, the community had grown to a population of 800. It had the Mountain City Hotel (a tent hotel), a log theater, the area's first post office, many other businesses and several saloons — in fact, the first saloon in the Rocky Mountains, it is said. Soon there were two newspapers and a Masonic Temple.

The Old National Theatre opened in the summer of 1862. The log structure was rechristened as the Montana Theatre on December 26th of that same year. Many musicals and Shakespearean dramas graced the stage, as well as plays with sociological overtones such as Rip Van Winkle and Uncle Tom's Cabin. Fire in 1873 closed the Montana Theater. Shortly thereafter, a new theater was established in Central City.

George W. Harrison, owner of The Old National Theater, was a popular man about town. One day before the grand opening on July 31, 1862, Harrison shot Central City boxer Charlie Swits to death as he departed from the Barnes Pool Hall. In the courtroom, Harrison's defense was based on the fact that he didn't like Swits. The jury didn't like Swits either — but they did like Harrison. The verdict was an acquittal.

As Central City and Black Hawk sprang up on each side, Mountain City was absorbed by both. In October 1869, the post office was officially moved to Central City. Because of the monumental impact of John Gregory's discovery, the area once known as Mountain City will remain a spot that changed Colorado history. Today several stone retaining walls, several old buildings, and the monument mark the site.

The barber shop of Jones and Urich, located in the basement of the Teller House advertised: "Three chairs, the largest and best in the country. Hot and cold showers, fifty cents day and night."

CENTRAL CITY
Location: 1 mile west of Black Hawk

Within weeks after John Gregory discovered gold on May 6, 1859, thousands of people flocked to the hills and creeks around Gregory Gulch. Famed New York newspaper man Horace Greeley, awed by all the commotion, decided he should see for himself and traveled to Colorado Territory. Although the miners were unimpressed with Greeley's lecture against drinking and gambling, he was, conversely, impressed with their discoveries. He declared, "...the news of your rich discovery shall go forth all over the world." And it did. As his articles circulated, people came from all over America and many countries abroad.

Although Mountain City was the first settlement in the area, Central City grew faster and quickly absorbed it, as it did most of the other camps in the vicinity. Central City, dubbed "the richest square mile on earth," was also Colorado's first major boom city and rivaled Denver for several years.

Gilpin County, which includes Central City, has produced nearly a half billion dollars in precious metals.

Possibly of more lasting importance is the culture which grew in an otherwise uncultured area. After the Montana Theater in Mountain City was lost to fire, the Belvedere Theater opened on the second floor of the Armory Building in Central City. Jack S. Langrishe developed local talent, imported additional talent, and staged plays to full houses. Due to the enormous success of the Belvedere, it was decided that funds should be raised to build a larger and more elaborate theater building. The Central City Opera House was born. Opening night, March 4, 1878, sold out at the rate of one dollar per reserve seat and seventy five cents per gallery seat. Central City became the entertainment capital of the mountains. Theater lovers saw many greats perform. Edwin Booth, Buffalo Bill, Oscar Wilde, and others graced the stage. The opera house closed in 1908. The population had decreased and economic times were tough. The theater reopened on July 16, 1932 to a performance of "Camille" starring Lillian Gish. Other top stars such as Faye Emerson, Helen Hayes, Shirley Booth, and Beverly Sills, came to Central City once again.

The Teller House, a lavish and exclusive hotel, was opened in 1872 by Henry M. Teller. Each room cost two dollars per night — an outlandish rate when the average hotel room usually cost about fifty cents. But, oh what amenities! There was a patented safety lock in each room that could not be unlocked from the outside, providing safety for the guests; running water on each floor; and a dining room that prepared exquisite cuisine. The Teller House became a rendezvous spot for society. Among its distinguished guests was President Grant who visited in 1873. Silverbricks worth several thousand dollars were shipped from the Caribou Mine and placed in the sidewalk at the hotel entrance for the occasion. On the floor in the Teller House Bar is the famous Herndon Davis painting "The Face on the Barroom Floor." Henry Teller served in the United States Senate a total of 29 years.

Although classes were first held in 1862, the Central City school building was not completed until 1870. There were 214 students registered for classes at the new school. The first public school library in Colorado, with nearly 2,000 volumes, was located here. A private school was constructed on Gunnell Hill in 1873. St. Aloysius Academy was built by the Sisters of Charity. A staircase of over 100 steps led to the front of the school.

Father Joseph P. Machebeuf held Catholic services most anywhere he could find space, including a billiard room. One Sunday, frustrated with the situation, he locked the doors and wouldn't allow his congregation to leave until they had pledged enough money to build a new Catholic Church.

For the satisfaction of the thousands of young single men, Central City had many brothels. One of the nicest was run by Ada Branch whose nickname was Big Swede. Elegant Ruby Lee seemed to be able to get her hands on big bucks. Madame B-Lizzie Warwick's brothel was another popular spot. Madame Lou Bunch, whose brothel was located near Warwick's is honored during the third weekend of each June with bed races and a Madame's Ball.

Pat Casey, an unmarried Irishman, loved the brothel houses, and frequented them regularly. He was shunned by the town ladies and frowned upon for his activities. Casey struck gold while digging a grave and became a wealthy man — he even had a street named after him. Legend has it that Casey would rent a fancy carriage on Sundays, pick up the "Hurdy Gurdy Girls," and ride past the community churches just so he could see the town ladies in a huff. Casey was the object of so much legend that a play was written about him, "Pat Casey's Night Hands," which was presented by Jack Langrishe's players in 1863.

A special story is that of "Aunt" Clara Brown, as she was known to the people of Central City. Clara was a southern slave who had bought her own freedom for one hundred dollars. She arrived in Central City in 1859 and opened up a laundry. According to Clara, "As soon as I had a sign painted saying laundry, I had customers." She would save a little cash and then grubstake a few miners. She used her money to help bring twenty-six slaves to Colorado and freedom. There is a stained glass window of Clara Brown in the state capital at Denver with an inscription that reads, "An ex-slave, who grubstaked miners and later led wagon trains of blacks from Kentucky to Colorado." Clara Brown was a black pioneer woman whose generosity left its mark on Colorado history.

As was the case with most boom towns, Central City had gold and other things on its mind — but not fire. When a fire started in the downtown business district at 10:00 a.m. on May 21, 1874, the city was totally unprepared. The conflagration was accidently started by a Chinaman who was burning incense on a bed of coals in an effort to exorcise an evil spirit. It burned out of control and by 5:00 p.m. most buildings in the downtown area were destroyed. Only one of the buildings that burned had fire insurance. It was the only house left standing on Pine Street. The Teller House and five other buildings were spared. Citizens immediately rebuilt the town with stone and brick. Determined that such a disaster would never happen again, Central City gave birth to its Rough and Ready Hook and Ladder Team, an award winning group of firefighters.

Most of the structures have undergone a complete renovation recently. Limited stakes gambling arrived in Central City October 1, 1991, and the community is experiencing an enormous new boom. In order to control the construction explosion, the city had to establish a moratorium on new building permits. As of this writing Central City has a population of 325, but over 3,000 are employed by the gaming industry. Tourists are flocking in to the many posh new casinos. Central City is, and always will be, a city steeped in colorful history.

The Colorado Central narrow gauge railroad had two engines. When one of the engines splashed into Clear Creek, the company went broke trying to fish it out.

BLACK HAWK
Location: 1 mile east of Central City

After the Gregory strike of May 1859, thousands of prospectors swarmed into the Gregory Gulch area. Tents and crude cabins sprang up everywhere. Many of the better strikes that followed were made near Black Hawk, such as the celebrated Bobtail lode. Those who did strike it rich were enough to keep others grubbing away at what an eastern newspaper called, "the world's deepest underground lunatic asylum." A town was platted and named after the famous Indian Chief.

A smelter, the Boston & Colorado, was established by Nathaniel P. Hill. Its advanced milling methods made it a great success. Ores were shipped to Black Hawk for processing from great distances.

Nearby, the Hendrie Brothers built the first mining machinery foundry in Colorado. It is believed that the cemetery on Dory Hill was the first in the state. Black Hawk also boasted one of the earliest permanent churches which was constructed in 1863. The Lace House, a Victorian home dripping with gingerbread, was also built in 1863. It is a tourist attraction today.

Black Hawk claimed more than 2,000 residents during the 1860s and possibly would have surpassed rival Central City in growth had it not been for its natural boundaries. Growth could only expand so far through the narrow canyon.

By 1865 there were thirteen saloons and gambling dens, many brothels, even a brewery — and Black Hawk was booming. There was always much activity around town. Ice shows at the local skating rink, and wrestling contests were commonplace.

Elizabeth McCourt Doe, known as "Baby Doe", and her husband, William Harvey Doe, Jr., once lived over a Black Hawk store. Disillusioned with her life, the legendary Baby Doe left Harvey and moved to Leadville where she found fortune and fame.

Today, Black Hawk booms again — this time because of gambling. On October 1, 1992 doors opened to casinos in Black Hawk, Central City and Cripple Creek and the renewed tourist trade is phenomenal. More than $763,000,000 flowed through gaming devices and across gaming tables during the first seven months of limited stakes gambling in the three communities.

"Wherever there was a rumor and a hole in the ground they built a camp."

NEVADAVILLE

Location: 1 mile southwest of Central City

Nevadaville was one of the towns in the Central City area which grew during the aftermath of the Gregory strike. Once known as Bald Mountain, Nevada, and Nevada City, the name was changed to Nevadaville to avoid confusion within the United States postal department. The community was segmented ethnically, with the Irish, Cornish, and Eastern Americans each having their own sectors of town. The Cornish and Irish were continually feuding.

Sports were very popular. The Cornish element organized the Mountain Daisy Cricket Club. Members of the cricket team were so confident of their abilities that they offered a fine silver pitcher to any team who could defeat them. A baseball club was also organized in Nevadaville. There was much activity on the two mile long snow run between Nevadaville and Black Hawk. The course was popular for sleds and snowshoes (which were similar to our present-day skis).

Fraternal organizations were prevalent in Nevadaville. The Masons built Colorado's first Masonic Lodge. Chapters of the Odd Fellows, the Red Men, and the Foresters followed.

During the 1860s, Nevadaville had a population of over 1,000. Residents packed up and fled during the panic of 1893, but as new activity flourished in the late 1890s, the population once again peaked over 1,000. Some of the town remains today, although most of it has withered away.

Mining laws, drawn up in 1859 for the newly created Russell District, were quite lenient. Section 67: Discovery Claims, holds that "Females have the same right as males. Youths under the age of 10 shall not be allowed to hold claims."

RUSSELL GULCH

Location: 3 miles southwest of Central City

William Green Russell, a seasoned prospector who had mined in Georgia and also California, located an abundance of free gold in 1859 in the gulch which carries his name. News spread fast and the rush was on. By fall, over 800 men in Russell Gulch were producing an average of $35,000 per week.

Population in the gulch swelled to 2,500 the following year. Placer mining was everywhere. Boarding houses, saloons, general stores, and scores of cabins sprang up throughout the gulch. So much digging had been done within the next three years, that the quality and quantity of ore drastically decreased. Most of the placer mining was abandoned, and the population of Russell Gulch diminished. A little mining activity has continued over the years. During prohibition days, bootleggers used many of the deserted mine shafts in Russell Gulch as warehouses for their merchandise.

There are about 200 residents in Russell Gulch today. Public buildings were converted to residences. Even the Odd Fellows Hall (which was a post office for a while) was converted into the residence of (they say in jest) the town's odd fellow.

PERIGO
Location: 4 miles southwest of Rollinsville

During the winter of 1859, a prospector named A. D. Gambell panned eight dollars of gold dust out from beneath the snow. He built a sluice and collected another ninety dollars worth of gold. After traveling to Denver for new supplies, during his return trip he was confronted by a group of men who had been following him. The men, who obviously watched him purchase his supplies with the gold he found, offered Gambell a choice, "Either be hanged here and now, or lead us to your strike." He agreed with the latter. In Gambell (or Gamble) Gulch they established the mining camp of Perigo.

The Perigo Mine and the Gold Dirt (around which a neighboring camp developed) were the best producing mines in the Independent Mining District. A 30-stamp mill was constructed. The mill was unusual because extra heavy pestles were installed to treat the low-grade ore found in the area.

In an extraordinary billiard match lasting 33 hours, John Q. A. Rollins won $11,000 from banker Charles Cook at a Denver billiard parlor in 1866.

ROLLINSVILLE
Location: 13 miles north of Black Hawk on State Highway 119

General John Quincy Adams Rollins built and owned the town of Rollinsville. Rollins Pass, the first pass over the Continental Divide into Middle Park, was operated by Rollins as a toll road. He speculated that everyone traveling through Middle Park would have to do so over his road and through his town.

Rollinsville was a "clean" community. Saloons, dance halls, gambling dens, and parlor houses were prohibited during the early history of the town. J. Q. A. Rollins liked to gamble, but did so elsewhere.

Rollins was a wealthy man who owned many mining properties, cattle ranches, and hotels. He operated the Rollins House for travelers visiting his town. He was also in the lumber business and owned a stagecoach line.

Among the many mining properties of the Rollins Gold Mining Company were the rich Gold Dirt and Perigo lodes. The most productive mines were the Ophir, Crown Point, Savage, Colorado, Comstock, and the New York.

The Rollins Steam Quartz Mill was located at Rollinsville on a downgrade from the mines. According to Fossett's *Colorado*, "It is terrace built, 100 feet long by 65 feet wide, and has one of the finest water-powers in the State ... "

Although it was an important community, Rollinsville never had a population over 200. Eventually saloons were allowed, but the importance of the toll road over Rollins Pass declined, as did the mining production. The town lives on at the intersection of State Highway 119 and the road to Tolland and East Portal.

WIDE AWAKE
Location: 6 miles north of Black Hawk via State Highway 119 and County Road 15

Articles have recently appeared in both the *Denver Post* and *Rocky Mountain News* regarding the ghost town of Wide Awake. It's for sale, and the asking price is $250,000. Thirty-four of the two hundred plus acres consist of patented and deeded mining claims. The balance is leased Federal land.

The camp was established shortly after its neighbor, Perigo, in the early 1860s, following the discovery of gold along Missouri Creek. Although gold was found no deeper than 400 feet at the Caledonia Mine it was the best producer in the immediate vicinity.

A couple of dilapidated buildings (one in the townsite, another nearby) stand amidst several piles of rubble. These and the foundation of the twenty-stamp Douglas Mill are about all that remain of Wide Awake. The town died when mining was discontinued during the 1920s. One resident hung on long after everyone else had gone. Wallace Stevens died in 1964 at age 97. Before he passed away Stevens told about twenty sacks of gold which were stolen from a nearby camp, brought to Wide Awake and buried for safe keeping. To this day the gold has never been found.

Who knows — if someone were to purchase this property, the camp which for years has been fast asleep could once again become "wide awake".

BALTIMORE
Location: 5 miles west of Rollinsville

Gold was discovered in the 1880s in Black Canyon, Jenny Lind Gulch, and on Baltimore Ridge. Many other discoveries followed and the area began to boom. The settlement of Black Canyon served the miners and mines in the

Phoenix District. The name of the town was changed in 1896 to Baltimore.

A post office was established that same year, and John Hatfield was named Postmaster. Hatfield was quite instrumental in the development of Baltimore. He owned many mining claims and was the original owner of the town site itself.

Amidst many cabins, a dance hall, drugstore, and saloon — the famous Baltimore Club — were constructed. Because the sale of liquor was prohibited in nearby Tolland (Mammoth), the Baltimore Club flourished. And it did so right on into the days of prohibition, because Hatfield simply ignored the law. When law enforcement officers carried Hatfield and his whiskey to Central City, the evidence quickly evaporated as it had a wondrous taste and the officers had to release Hatfield for "lack of evidence."

In 1905, John Hatfield married an attractive, cultured, and artistic lady, Lillian Woodard. She was a composer and an accomplished pianist who had studied at the Boston Conservatory of Music. Soon a piano arrived in Baltimore — later, a wagonload of curtains. The dance hall was converted into an opera house. Much entertaining was done during the ensuing years. In fact, Lillian Hatfield may have gone overboard "entertaining." It is rumored that the dressing rooms on the second floor of the opera house were converted into a brothel.

Lillian's piano still sits inside the log house "Sunbeam," once the Hatfield home. Directly across the street the Baltimore Club stands precariously — propped by large timbers. Several cabins have been renovated and are inhabited today.

"Some friends are friends only up to their pockets."

APEX

Location: 7 miles northwest of Central City via Forest Route 176

Although it is located only a few miles from the hub of earlier mining activity in the Central City area, Apex wasn't founded until thirty-five years later and experienced the height of its boom in 1899 and 1900. Possibly because of its late start, Apex grew as a very sophisticated community. It was the center of the late blooming Pine Creek Mining District.

The Apex Hotel was popular and had far reaching appeal. The large Palace Dancehall was always busy and prosperous. Among the town's many buildings were two churches, a schoolhouse, and a newspaper, the *Apex Pine Cone*. The population peaked at over 1,000.

There were many mines within a short distance of Apex which produced well. The best was the Mackey Mine, named for Dick Mackey who staked the original claim. The mine was purchased by a prospector named Mountz and his partners. With the mine offering a modest yield, the partners absconded with all of the capital leaving Mountz a meager $400.00 with which to work the mine. His money quickly ran out. Totally disgusted by the course of events, Mountz decided to blow the tunnel to smithereens. He planted dynamite. The ensuing blast revealed a magnificent lode. The ore assayed at $1,800.00 per ton. Mountz was rich.

A few of the remaining cabins in and around Apex are still occupied today.

AMERICAN CITY

Location: 9 miles northwest of Central City via Forest Route 176

About a mile and a half from Apex, nestled on the wooded side of Colorado Mountain, lies American City. It was a mining and milling town that revolved around the activity of the Boston-Occidental Mining Company.

Gus Meyer, a master craftsman from Germany, constructed the mill in 1903. A school was built which not only served American City, but also the community of Nugget, two miles below. A hotel followed, but the structure was not befitting its elegant name, the Hotel Del Monte. The town, which never grew very large, was reliant on Apex for commerce, merchandise, its post office, and newspaper.

GILPIN

Location: 6 miles south of Rollinsville via State Highway 119 and County Road 12

Gilpin, which is located in Lump Gulch, sprang up at about the same time as it's neighbors Perigo and Gold Dirt in adjacent Gambell Gulch.

Fairly rich strikes were made at the War Eagle and the Gettysburg. There were several lesser mines in the vicinity also, such as the Victoria. Ores were shipped out in two directions for processing. Some went to the mill at Perigo, others to Rollinsville.

The town was named in honor of Colorado's first territorial governor, William Gilpin. The settlement was never large but did have a general merchandise store, boarding house, post office, blacksmith shop, schoolhouse, and a baseball team.

The mines lacked longevity and newer strikes failed to pan out — so Gilpin declined. Today a couple of the original buildings are mixed in with summer homes.

Mountain City was established near the Gregory Lode. *Colorado Historical Society.*

Central City in 1864. *Colorado Historical Society.*

Above the bank of Warren Hussey and Co. at Central City was the Concert Hall, a gambling palace operated by Barnes and Jones. *Colorado Historical Society.*

Shaft buildings on the Gregory vein climb the hill, at left, in this 1864 photo of Black Hawk. *Colorado Historical Society.*

GILPIN COUNTY 27

Silver bars stacked at Hill's Smelter, Black Hawk, about 1872. *Colorado Historical Society.*

Nevadaville about 1890. *Colorado Historical Society.*

28 NORTH CENTRAL REGION

Early Russell Gulch. *Denver Public Library, Western History Department.*

Looking west through Perigo in 1899. *Denver Public Library, Western History Department.*

The Gooch Hotel (building with the veranda) in Rollinsville was destroyed by fire. At left is the Perigo Mine's office. *Denver Public Library, Western History Department.*

Apex about 1890. *Colorado Historical Society.*

30 NORTH CENTRAL REGION

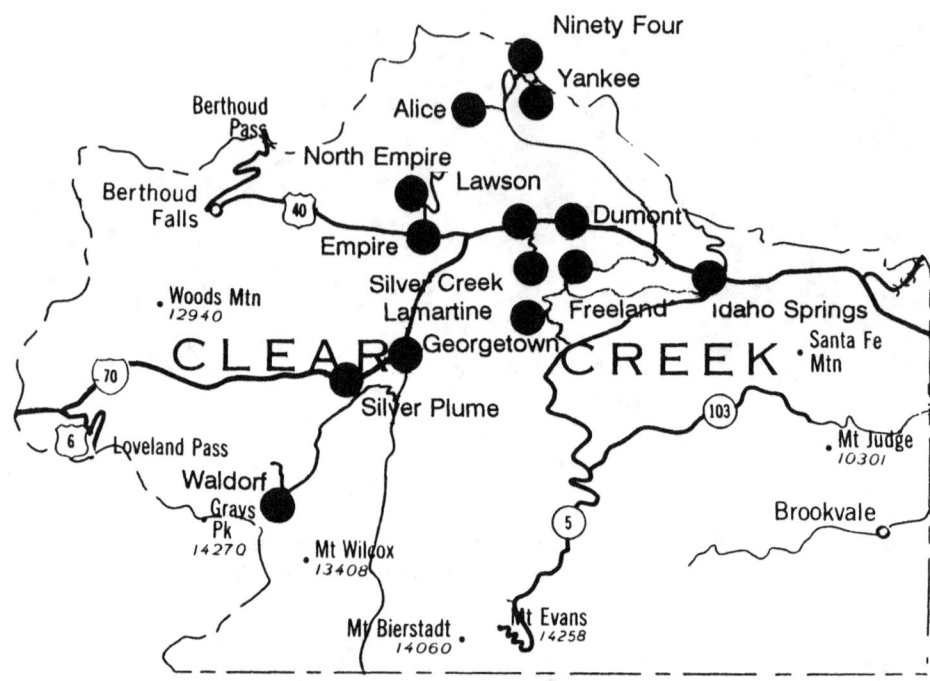

CLEAR CREEK COUNTY

The diary of George A. Jackson held the following entry for January 7, 1859: "Clear day. Removed fire embers and dug into rim on bedrock. Panned out eight treaty cups and found nothing but fine colors. Ninth cup I got one nugget of coarse gold. Feel good tonight."

IDAHO SPRINGS

Location: 11 miles east of Georgetown on I-70

George Andrew Jackson, a native of Missouri, discovered gold in January of 1859 at a spot to be known as Jackson Bar on Chicago Creek. A few days later he hiked back to meet his friend Tom Golden at Arapahoe (now the city of Golden). During the ensuing weeks Jackson was joined by a group of 22 prospectors from Chicago. The group packed up supplies and on March 7th departed for the location of Jackson's discovery. During the first week the group panned out $1,900 in gold. The rush was on. Hundreds of prospectors poured into the valley. Chicago Creek was just one of many tributaries that dumped into Clear Creek, so prospectors spread out to comb the nearby hills. Edward's Placer was located on Chicago Creek. More discoveries followed along Clear Creek at Payne's Bar, Spanish Bar, and Idaho Bar (where the business district of Idaho Springs would soon be located).

The Colorado Territory was created in 1861 and Idaho Bar was named seat of the newly established Clear Creek County (Georgetown became county seat in 1868). By this time forty substantial houses had been constructed, as had the hotel of F.W. Bebee — the Bebee House. Dry goods and grocery stores followed — as did a blacksmith shop. The settlement was known as Idaho Bar until 1873 when the town of Idaho Springs was officially organized. The manufacture of bricks began about this time by the Crosson family. Most of the commercial buildings subsequently erected were of material from the local brickyard. The Hotel de Paris was originally a two-story wooden structure. Its owner Michael B. Graeff became increasingly concerned with the danger of fire and rebuilt the hotel with brick in June 1880. Two months later it burned. After renovation, the building reopened as the Queen Hotel. In 1881 the elegant Club Hotel was built of fine Victorian architecture. Restaurant, saloons, and variety shops of all types appeared throughout the business district. A bank was built, then another, and a Masonic Lodge as well. The Presbyterian Church was the first of several places of worship.

One of the most notable landmarks at Idaho Springs is the large waterwheel adjacent to I-70. The wheel was constructed by Charles Tayler (spelled with an "e") to power a small stamp mill on Ute Creek during the 1890s. It was relocated to its present site, below Bridal Veil Falls, in 1945. In 1988 the Tayler Waterwheel received a new foundation and extensive restoration.

As mining activity increased, Idaho Springs became a major smelting center. There were over two dozen treatment plants of one type or another by 1903. In 1892 the 22,000 foot Argo Tunnel (originally called the Newhouse Tunnel) was bored from the Argo Mill in Idaho Springs through the hills to Central City. The cost of construction was $10,000,000, and tapped several mines allowing expedient access to the railroad. Labor strife hit Idaho Springs in 1903 when miners struck for an eight-hour day. Some mine buildings were dynamited during the violence that followed. Things finally settled back to normal when the union leaders were run out of town.

Traveling from Idaho Springs, through Virginia Canyon, to Central City was such a harrowing experience that the trail was dubbed "Oh My God Road". It was one of the first roads through the mountains. Another road followed Fall River, then crossed Yankee Hill into Central City. Eventually other roads connected with Georgetown and other places. The narrow-gauge railroad was completed through Idaho Springs and on to Georgetown by 1877.

H.A.W. and Augusta Tabor were among the town's earlier residents. A distinguished visitor was Ulysses S. Grant who stopped while touring gold camps in 1873. Mark Twain and Doc Holliday were among others who passed through the community. Buffalo Bill Cody died shortly after visiting Idaho Springs. It is said, he had his last drink at the Duck Inn - a whiskey and cider mix.

The population of Idaho Springs is swelled throughout the year by the influx of tourists. The community is finally doing an adequate job of promoting its rich history.

Mill City (Dumont) had the first sewing machine in the territory of Colorado — it is believed.

DUMONT

Location: 5 miles west of Idaho Springs via I-70

John M. Dumont, a native New Yorker, headed for the Colorado hills in the spring of 1859 to search for gold. He made two promising strikes but had no money to develop the lodes. Being a man of ingenuity, Dumont hired miners who were hungry and down on their luck. To get things going he paid them with groceries which he had purchased on credit. The mines began to pay dividends.

A settlement originally named Mill City cropped up in 1859 near one of Dumont's properties — the Freeland Mine. The camp was named for the ore crushing arastras established there.

An elaborate hotel, the Mill City House, was constructed in 1868. Fine furnishings were shipped in from the east. The Mill City House was also a stage station, toll house, and the center of activity in the community. The upper floor was used as a meeting hall, theater, and opera house. The establishment also boasted a posh billiard room and bar.

Edgar Freeman constructed a stamp mill. A general store was also built, as were many cabins, but the anticipated boom never came. Dumont had ownership in several mining properties in the vicinity — the Whale, Lincoln, Lone Tree, and the aforementioned Freeland. Mill City received a boost in 1879 when Dumont sold the Freeland to a Californian for $250,000. The sale brought new money into the community.

The *Georgetown Courier* stated, in June 1880, "Dumont is a new name that has been applied to Mill City in honor of John M. Dumont of Spanish Bar." The town's main street was named Dumont Avenue as well. Another hotel, the Unadilla House, and the saloon of Joshua Kramer were built opposite each other on Dumont Avenue.

Shortly after the name change, Dumont's town scribe recorded the following (in November 1880): "...I found that Patrick Murray had been shot in the left shoulder by James Brown. It appears that Murray had been imbibing freely that evening and upon returning to his cabin, found Brown in bed. Murray had a knife in his hand, and Brown, losing control of his temper, pulled his pistol and fired. The cause of the quarrel emanated from an old grudge. 'Brown has skipped,' and Murray still carries the lead." In a separate incident the town scribe also recorded, "A miner, whose name I did not learn, employed on the Lone Tree Mine at Freeland, while warming a stick of giant powder over a candle lost one of his hands to the wrist by its explosion on Monday."

In 1966, a fire burned down the firehouse. Today, the community tries to coexist with the hustle and bustle of Interstate 70.

The first woman to arrive in Empire was the sheriff's wife, Mrs. James Ross, who was presented a free building lot to commemorate the occasion.

EMPIRE

Location: 9 miles west of Idaho Springs via I-70

In the fall of 1859, toward the eastern edge of the valley that would eventually become the townsite of Empire, Dr. Richard Bard constructed the first log cabin in the vicinity. George Merrill and Joseph Musser soon built a second cabin across the meadow to the west. Many tents sprang up between the cabins, and a settlement began to take shape. Realizing a need for self-government, a meeting was held in August 1860, and the Union District (Union Mining District) and the Valley City Company were organized. Mining claims such as the Iowa, Buckeye, Eureka, and Wisconsin were quickly recorded. The townsite, which was briefly known as Valley City, was staked.

By early the following year, the community had a blacksmith shop, Frank L. Andre's Hardware (onto which the post office was attached), and dozens of cabins. A two-room log cabin originally served as town hall, courthouse, and sheriff's office. On February 5, 1861, the following was documented by David J. Ball, town recorder: "Be it known, the Methodist Episcopal Church claims by preemption for church purposes 150 x 200 feet of ground being the east portion of Block 30 in Empire City." This was the first recorded use of the name Empire City. The town was named by Henry DeWitt Clinton Cowles for his Empire State — New York. Eventually the name was simply changed to Empire.

Early Empire had one hotel, Charles Bermister's Park House, where lodgers complained that the scanty hay stuffing their sacks was poor protection from the earthen floor. There was a schoolhouse which was no more than a small log cabin. Although Andre's store came first, it wasn't long before a combined dry goods and grocery opened — the James Peck & Company. Dean W. King constructed the first two-story log house in the settlement.

Church services were originally held in a tiny chapel constructed of rough-hewn logs. The aforementioned Methodist Episcopal Church was never constructed. An Episcopal church was built, however. The Methodists worshipped in the new Town Hall once it was finished.

Some of the other buildings constructed were a livery stable, brewery, meat market, and three saloons — two of which had billiard tables. One of the bars was owned by a fellow named Mix — who did just that. Under pressure from a temperance movement, each of the bars added a reading room to add a little culture and refinement to their establishments.

Many gold mines changed hands in 1863. The Civil War had depleted the workforce and shot prices upward for materials and supplies. The cost of mining became more expensive. At the same time, however, the value of gold escalated to $50 per ounce. Certainly there was a new demand for gold mines and mining stock. Struggling prospectors seized the opportunity to sell their prospect holds. Judge Henry DeWitt Clinton Cowles and George L. Nickolls opened the Empire City Mining Agency, a partnership formed exclusively to deal in mining and mill properties. The lively boom was short, however, and by the spring of 1864 the gold market began to collapse. Mines had exhausted high-grade surface ores and had reached pyrite, difficult to crush and with layers where free gold couldn't be found.

John E. Leeper, J. H. Coombs, and George L. Nickolls had each constructed separate stamp mills. In 1865 ore could be treated by the stamp-mill process for about $30 per ton. On an average, mills saved 20% of the assay. A rich ore which assayed at $200 per ton might yield $40 — a profit of only $10 per ton.

Killings were not uncommon in the early west, and one occurred early September 20, 1864, in the town of Empire. Apparently the oxen belonging to Gallant V. Hunter had been annoyed for some time by the dogs of Peter Geary. They must have annoyed Hunter also, for he shot one and vowed to get the other. In retaliation Geary chased down Hunter's dog and killed it. Outraged, Hunter ran after Geary and shot him to death in front of the Town Hall. Hunter was charged with manslaughter, faced trial, and was acquitted because he had "reasonable grounds to fear for his own life." He was defended by noted attorney, Henry M. Teller, of Central City.

During the feud between the Pelican and Dives mines in Silver Plume (both were tapping the same vein), Jacob Snyder, a Pelican owner, was shot to death by a mercenary named Jack Bishop. After the murder, Bishop fled to Empire to seek out his friend, Harry Carns, who lived at the brewery of Paul Lindstrom. He told Carns that a posse was close behind and asked to be hidden. Bishop was led into a dark cellar and hid behind some large brewing vats. His horse had been hidden in a nearby thicket. When the sheriff arrived, he asked Carns if he had seen Bishop. Carns answered, "Sure, he's hiding behind old man Lindstrom's beer kegs. Go in and look for him." The sheriff didn't believe Carns, mounted, and rode off with the posse. Mrs. Lindstrom furnished Carns with food and blankets but was not given a reason for the supplies. Several months later Mrs. Lindstrom received an envelope containing money but no letter or note. She had assumed the food and blankets were for Jack Bishop ... and now she was certain that the money had been sent by him as payment and thanks.

Upon its completion, the new Town Hall became the center of social activity and entertainment. Somebody came up with a great idea for recreation — indoor croquet. Small holes were bored in the floor of the Town Hall for placement of wickets. After days of hard work, townspeople gathered in the evenings to participate or cheer on their friends. Because the terrain at Empire was hilly, uneven and rocky the indoor sport was a natural and very popular.

James and Mary Grace Peck, two of Empire's early residents, loved to entertain and did so as often as possible. In 1872 they carried their hospitality one step further and converted their home into a hotel — the Peck House. Mrs. Peck served the finest meals in town. A favorite dish was mountain trout with a side of wild mountain berries.

Empire was not incorporated until 1882, long after its population had dwindled. Only 39 miners took leave from work to cast their vote for the incorporation.

The people of Empire persevered — and did so right through the depression of 1929. When the price of gold jumped nearly $15 per ounce to approximately $35, the economy of Empire took a sharp turn upward. Gold mines were reopened, mills were refurbished, and many of the townspeople began driving shiny new automobiles. In 1935 several mines merged and incorporated as the Minnesota Mines. The company built a 100-ton cyanide plant, then a concentration works. They operated three shifts per day, seven days a week, and processed over 200 tons of ore each day. After the outbreak of World War II when gold-mining was declared a non-essential industry by the government, they were prohibited from obtaining supplies necessary for mining operations such as powder, steel, and machinery. Minnesota Mines and other gold-mining properties totally shut down.

On July 3, 1938, a caravan of motorists, driving the new U.S. Highway 40 from coast to coast, paraded through the heart of Empire and proceeded to the summit of nearby Berthoud Pass amid much celebration. Thousands participated in the festivities or joined the motorcade which was led by Colorado Governor Teller Ammons.

A marvelous transformation occurred in the mining industry when Dr. Alfred Bernhard Nobel created dynamite in 1866. Dr. Nobel (of Nobel Prize fame) had also discovered nitro-glycerine. Dynamite, known then as "giant", first appeared in Clear Creek County in 1870.

NORTH EMPIRE

Location: 1 mile north of Empire

Wire gold was discovered on Eureka Mountain by Henry DeWitt Clinton Cowles and Edgar Freeman in the fall of 1860. They established the Nebraska City, Uncle Sam, and Utah claims. Throngs of prospectors quickly poured into the area. Claims were staked everywhere. Placer mining on Silver Mountain yielded gold so rich that part of the mountain soon looked as though it had been "skinned alive." The settlement which sprang to life on the side of Covode Mountain, opposite the Silver Mountain mining properties, was named North Empire in 1861. For some time it was the center of vigor, vivacity, and just plain hard work. The community, often considered a "suburb" of Empire, had two boardinghouses, a general merchandise store, and a few saloons during its early days.

Between the forks of Lion Creek (sometimes spelled Lyon), Henry DeWitt Clinton Cowles also discovered the first true fissure silver lode in Colorado and named it the Ida Silver. Silver Mountain was named for this strike, but ironically the mountain yielded gold and only a touch of silver.

North Empire (often referred to as Upper Empire) boasted many productive gold mines. An important property was the Silver Mountain Mine. It was an early discovery which was sold by the enterprising W. H. Russell to the Star Gold Mining Company of New York in 1863. Some of the productive properties on Covode Mountain were the Neath, Cashier, Cambridge, Covode, Big Chief, Pilot, and the Paymaster. The Peck Gold Mining Company (of James Peck) operated the Gold Dirt and the Equator. Other good properties were the Tenth Legion (one of the richest strikes in the Union Mining District), Conqueror, Pittsburgh (recorded as Pittsburg), Gold Fissure, Pioneer, Empire, Comet, Benton, and the Atlantic (later to become an important property of the Minnesota Mines Company). John Dumont subsequently purchased the Neath, Pioneer, and Benton (near which he built his Blue Boarding House). The Conqueror Tunnel facilitated access to the Patsey and other lodes which it crossed.

Hydraulic mining arrived in 1878. Timothy G. Negus, Frank M. Taylor and Robert S. Morrison organized the Empire Ditch and Placer Company with capital of $400,000.

The company which owned about 40 acres on the south slope of Silver Mountain began to "wash" down 25,000 cubic yards of earth per week. A few years later Judge R. E. Rombauer and Amos Morse placer mined the Coupon and Pauline with hydraulics.

Mining had its ups and downs in North Empire. During the early 1900s when many of the mines were closed, the Gold Dirt was producing adequately. Investment capital was pumped into the Conqueror Mining and Milling Company in 1910. A new mill was constructed, as was a boardinghouse with accommodations for fifty men. The Pioneer Mill was shipping regularly to the Denver Mint in 1913. The merger of several properties to form the Minnesota Mines Company in 1934 created another flurry of activity which continued until World War II, at which time the government closed down the gold-mining industry. Order L-208 not only stopped gold-mining but it also spelled doom for North Empire.

The U.S. Post Office at Waldorf, located at an altitude of 11,666 feet, was listed as the highest in the country.

WALDORF
Location: 5 miles southwest of Georgetown

Large deposits of silver sulfide were discovered in 1867, and about seventy-five claims were staked out on the slopes of Mount McClellan.

Waldorf grew as a mining settlement and a stagecoach stop on the road to Argentine Pass. The high altitude (11,666 feet) and heavy snows made it virtually impossible to transport ore during the winters. Originally, Waldorf was a "summer" camp and remained so for several years until E. J. Wilcox became involved.

Edward John Wilcox (1857-1928) was a mining man, railroad man, developer, and an ordained Methodist minister. Nearby Mount Wilcox is named for him. Wilcox was instrumental in the success of Waldorf.

Wilcox founded the Waldorf Mining and Milling Company (later, the Waldorf Consolidated Mining Company), which bought out many of the more productive claims in the area. In order to transport ores and supplies to and from Waldorf, Wilcox built the Argentine Central Railroad which connected the camp with Silver Plume. Because the narrow-guage had grades up to 10 degrees and thrilling curves to about 145 degrees, coupled with breathtaking scenery, it became quite a tourist attraction. Also, tourists could visit the ice cave, an abandoned tunnel in which marvelous ice crystals remained throughout the year.

Waldorf — which by then was a company town — boomed. There was a hotel, boardinghouse, post office, stables, stores, and cabins. A large mill, powerhouse, and machine shop were also located at the town. The Argentine Central was, at the top of the pass, the highest steam railroad in the world. It is said, that 106 peaks can be viewed from the top.

A sign was needed for one of Georgetown's new saloons. Trouble is, the saloon keeper, who couldn't read, didn't much like the only sign painter in town — and vice versa. Well, a sign was painted and hung that read: "We sell the worst whiskey, wine and cigars." The joke was on the painter, however, for the sign attracted so much attention that the saloon keeper did a landslide business.

GEORGETOWN

Location: 11 miles west of Idaho Springs on I-70

George Griffith, a native of Kentucky, discovered gold in June 1859 near the confluence of Clear Creek and Leavenworth Creek. He staked his claim, then hiked to the tent colony at Mountain City to find his brother David who was working the Gregory Diggings. George, David and three other prospectors returned to the claim and worked it until winter set in. By then they had taken out about $500 in gold. The brothers returned the following spring with their father, George's wife Elizabeth, and the rest of their group. They built a cabin and called the spot George's Town in honor of George Griffith. They also built a make-shift stamp mill and established the Griffith Mining District. Several other prospectors ventured into the area, but the hills and creeks yielded a disappointing amount of gold. The Griffiths could barely scratch out a living. Oddly enough, the Griffiths found large amounts of silver-bearing ore in their lode. They tried to melt the material to make bullets but were unsuccessful. So they labored through the seemingly unvaluable ore in an effort to follow their tiny vein of gold. Times were tough during those first years. The Griffiths had to climb the steep trail over Union Pass to pick up their mail from the post office at Empire City. By 1863 there were still only four log cabins at George's Town, but things were about to change.

In 1864 Robert Steele, James Huff, and Robert Layton discovered quartz containing silver and staked a claim which they called the August Belmont. This time the rush was on. Hundreds of claims were staked throughout the hills, and a new, more substantial settlement was established at the base of Leavenworth Mountain, and was named Elizabethtown for Elizabeth (wife of George). Elizabethtown and George's Town agreed to combine, and a post office was established in June 1866 in a log building at the intersection of Sixth and Taos Streets, under the settlement's new name — Georgetown. John Lafferty was the first postmaster.

Other silver mines were located such as the Anglo-Saxon, Paymaster, Seven-thirty, Payrock, and the Sunburst. The Pelican and Dives mines were perhaps the most famous. Both mines tapped the same vein. The dispute which arose ended in violence and murder. Eventually the two properties were combined and simply known as the Pelican-Dives (see Silver Plume for more of this story).

As a result of the silver explosion, Georgetown grew rapidly. 1867 was a year of much commotion — as carpenters were hurriedly framing up buildings everywhere; freight wagons and jack trains crowded the streets; merchants sold goods and supplies from their wagons and tents; and Georgetown became known as the "Silver Queen of the Rockies." During that year several hotels and boardinghouses opened, including the posh Barton House. McClellan Hall (not to be confused with the McClellan Opera House) was also built in 1867. The Ohio Bakery was also constructed (the building was moved and is presently the Georgetown Community Center). Houses, shops and saloons sprang up everywhere. The initial issue of the *Colorado Miner*, Georgetown's first newspaper established by Dr. J.E. Wharton and A.W. Barnard, hit the streets in 1867 as well.

In the years that followed, the community's growth continued. A second newspaper appeared — the *Georgetown Courier*. On the corner of Sixth and Taos Streets (a vacant lot today) the McClellan Opera House was constructed in 1869. Many traveling shows and top entertainers graced its stage before the two-story structure burned to the ground in 1892. The brick schoolhouse was constructed on Taos Street in 1874. During October of the following year the doors opened at the fashionable Hotel de Paris. Louis DuPuy, a frenchman, purchased the Delmonico Bakery then proceeded to build it into the famed hotel. The finished hotel actually consisted of three buildings covered with a new masonry facade. The luxurious hostelry boasted carved walnut furniture, tapestry drapes, Haviland china, a divided staircase, and a library. Exquisite cuisine was complimented by an array of fine wines. DuPuy's closest friends were Monsieur Galet, a French cabinet maker, and his wife. Upon her husband's death Sophie Galet moved into the Hotel de Paris, to help with its housekeeping. When Louis DuPuy died of pneumonia in 1900 he left the hotel to Madame Galet. The hotel is operated as a museum by the Colorado Chapter of the Society of Colonial Dames.

By 1876 Georgetown had become a substantial city. There were fourteen hotels and boardinghouses, five churches, a pharmacy, stores and shops of every variety, a school, bank, jailhouse, courthouse, two opera houses, a community hall, four hose companies (fire departments), a Masonic Lodge, two newspapers, a telegraph, water works, doctors and lawyers, mining and assay offices, mills, several brothels, and over thirty saloons. The following year a railroad depot was built to coincide with the arrival of the Colorado Central.

William and Priscilla Hamill were pillars in the community. Both were in the center of Georgetown society. Hamill became wealthy with mining investments such as the Pelican-Dives property at Silver Plume. He owned the Merchants National Bank which was housed in a group of buildings which he constructed in 1881 known as the Hamill Block. Among his diverse interests, William A. Hamill was a politician, rancher, mining magnate, newspaperman, road builder, banker, and general contractor. At the corner of Argentine and Third Streets stands the Hamill House. Part of the house was constructed in 1867 by Joseph

Watson (Hamill's brother-in-law), then acquired by William Hamill in 1874. As his wealth grew so did the house. Today, the house is owned by the Georgetown Society and is open to the public.

Although Georgetown was a refined city, it had its wild element as well. There were gambling dens, brothels, and saloons — in fact, there was one saloon for every 150 people. Most of the "action" was on Brownell Street which sported five fancy bordellos. Mattie Estes' parlor was plush and popular. Madame Mollie Dean was murdered by a jealous lover. Shortly after a miner was shot to death at her brothel, Madame Jenny Aiken burned to death when her parlorhouse went up in smoke. In a separate incident, a jealous squabble over a fancy lady resulted in a miner being shot to death.

Silver Plume, two miles west of Georgetown, was 638 feet higher in elevation. To obtain an incline of less than four percent it would be necessary for the Colorado Central Railroad to lay more than three miles of track between the communities — and it had to do so through the narrow passage known as Devil's Gate. A series of curves were engineered including a loop whereby the railroad crossed over itself on a 300 foot long trestle. The "Georgetown Loop" was finally completed, and trains rolled into Silver Plume in March 1884. The railroad was abandoned in 1939 by the Colorado & Southern (which then owned the railroad). The trestle was dismantled and rails were removed. History has been preserved, however, for the entire narrow-gauge route between Georgetown and Silver Plume has been reconstructed. The Georgetown Loop Railroad crosses the 95-foot high Devil's Gate High Bridge to the thrill of thousands of tourists each year.

Repeal of the Sherman Silver Purchase Act in 1893 rocked the silver industry. The population of Georgetown declined during the ensuing years to a low of 300 in the 1930s. With the community's revitalization its population has climbed to about 900.

"An old timer is a man who's had a lot of interesting experiences — some of them true."

SILVER PLUME

Location: 2 miles west of Georgetown on I-70

Silver mining blossomed in the late 1860s and early 1870s in Brown, Cherokee, and Willehan gulches. One-half mile west of the future town of Silver Plume, the camp of Brownsville sprang into existence. As mining activity increased, residents spread further into the valley and established the community of Silver Plume. It developed as a miner's town — of low and average income workers and their families. The people of Brownsville eventually moved to Silver Plume to escape the continual threat of mudslides. Those who were more affluent, including many of the mine owners, opted to establish residency nearby in more sophisticated Georgetown.

Ill, and seemingly on his death-bed, Owen Feenan confidentially advised two friends of a rich strike which he had been keeping a secret. Feenan recuperated to find that "his" Pelican Mine was being worked, and that he had been excluded from the operation.

Later, speculation that the Pelican and Dives mines were tapping the same vein created a feud which wound up in litigation. Twenty-three separate suits were on file at one time contesting rights to the rich eight-foot silver vein. Pelican owners believed that the Dives' management was stealing their ore. On one occasion, an accident supposedly claimed six lives. Coffins were lowered into the mine — then raised and hurried away. The accident was a hoax, and the coffins contained high-grade ore. Armed guards were hired and stationed at both mining properties. Jacob Snyder, one of the Pelican's owners, left the mine one morning in 1875 and departed for Georgetown. En route he was accosted by Jackson Bishop, a Dives' lessee. Bishop chased Snyder to the edge of town, fatally shot him through the head, then escaped. The two mining properties were ultimately purchased by magnate William A. Hamill, of Georgetown, and consolidated as the Pelican-Dives. Hamill, who had purchased the Dives for $50,000 at a sheriff's sale, sold the combined properties in 1880 for $5,000,000 to become one of the richest men in the region.

In addition to the Pelican-Dives, located in Cherokee Gulch, there were several other fine producers in the vicinity. The Payrock, Corry City, and Burleigh mines yielded high-grade silver, lead, zinc, copper, and some gold. The Terrible (another property of W.A. Hamill), Dunderberg, Smuggler, Mendota, and Seven-Thirty mines were located in Brown Gulch. Also, rock was shipped throughout parts of the state from the Hamill Granite Quarry located at the east end of the gulch.

Above Brown Gulch stands a monument which was erected in memory of Clifford Griffin who died by his own hand in June 1887. According to legend his fiance died on the eve of their wedding, and the young Englishman came to Colorado in order to take up mining and forget the past. He discovered the Seven-Thirty Mine on Columbia Mountain and built a cabin at the site. Many evenings the lonely and grieving miner would play his violin in front of his cabin, and his music would echo through the gulch. He left a note requesting that he be buried at that site, then took his own life. His request was granted.

Silver Plume incorporated in 1880 at which time its population was near 1,000. The community had a Catholic Church. The Methodist Church was originally located between Brownsville and Silver Plume. When Brownsville's population moved to Silver Plume, the church was relocated as well. A number of wood-frame businesses were densely packed into Silver Plume's commercial district. There were groceries, meat markets, dry goods stores, laundries, boardinghouses, billiard halls, and many saloons — several of which lined Main Street. Following the completion of the famous Georgetown Loop (see

Georgetown) the railroad arrived into Silver Plume amid much fanfare. The grand loop became an instant tourist attraction. Visitors flocked to Silver Plume and Georgetown from afar to ride the curves of the Union Pacific narrow-gauge.

In November 1884 a fire broke out during the night in a saloon on the east end of Main Street. The conflagration swept out of control through the wood-frame buildings of Silver Plume's commercial district. The citizens of Silver Plume went to work rebuilding their town.

Eugene Morgenthau lost his dry goods store during the fire of 1884. By the following year he had constructed a large new brick building and had resumed business. Another store which was destroyed — the Pioneer Merchandise Company — was succeeded by the general store of Mortiz Neuman and Matt O'Neill. Mary Thomas built a fine new grocery store to replace the one which burned down in the fire. After the offices of the *Silver Plume Colorado* were destroyed, William Blanton established the *Silver Plume Jack Rabbit*, a short-lived newspaper, and then published the popular weekly — the *Silver Plume Silver Standard*. Other citizens pitched in as well, and within a few short months much of the downtown area had been rebuilt.

The Terrible Silver Coronet Band was not so-named for the quality of the music they played but rather because they were sponsored by the Terrible Mine. The Payrock Mine also had a band — simply named the Payrock Band. These, and other bands, provided entertainment for the community. Band music was so popular that a gazebo-style structure was constructed in 1904 — the Silver Plume Bandstand.

Mining has continued off and on throughout the years. Part of the town was razed when Interstate 70 cut its path through the valley. The community, with its current population of 250, is host to thousands of tourists each year. Silver Plume looks much as it did a century ago.

LAWSON

Location: 7 miles west of Idaho Springs via I-70

On June 4, 1871 wagonmaster Alex Lawson transported the John Coburn family to Clear Creek County. Coburn established the settlement of Free America (later to become Downieville) and built a hotel there. Lawson and Coburn's daughter, Kate, fell in love and discussed marriage — something Coburn strongly opposed. Kate encouraged Lawson to become a more settled businessman and he did so by building his own hotel six miles from Georgetown. To the chagrin of Coburn, Kate and Alex eloped on August 7, 1873. Their Six Mile House became a popular hotel, and naturally a stopover for Lawson's stagecoach as well.

When the railroad arrived into Clear Creek Valley, Alex Lawson convinced its planners to establish a depot at the settlement. The town of Lawson was born. Another factor which contributed to the early success of Lawson was the gold mining activity which blossomed just over a mile north of town at Red Elephant. The Free America, Saint James, Young America, Dexter, and Lulu mines were all fine producers. Some miners lived at the little camp at Red Elephant which had a population of about 200. Many lived at Lawson which during its peak years had about 500 residents.

When Anson P. Stevens, Lawson's wealthiest man, first arrived, he immediately built a mill and concentrating works, and an attractive stone residence as well. Concentrating was a new process whereby tailings (previously discarded) could be treated for additional yield. Stevens was a shareholder in the Paymaster Mine on Covode Mountain at nearby Empire. Then he purchased the Cashier which was close to the Paymaster. Subsequently he leased the Cashier to three partners who in-turn had to ship their concentrating ore to the A.P. Stevens' Mill at Lawson. It was typical of the type of business deals which made Stevens a multi-millionaire. It is said that one year he couldn't think of a suitable Christmas gift for his wife, so he gave her a million dollars.

Much of Lawson disappeared when Interstate 70 cut a wide swathe through the valley. The stone house of Anson P. Stevens was among several fine homes to be demolished. That which remains is still inhabited.

At Silver Creek a vast underground chamber was excavated in the O'Connell Tunnel — large enough in fact to hold a dance. So they did — and among others, the entire newspaper staff of the Georgetown Miner *was invited to the subterranean affair.*

SILVER CREEK
Location: 2 miles south of Lawson on Silver Creek

There was mining activity near the future site of Silver Creek for several years prior to establishment of the camp. Evidently some miners were tired of the two mile climb from Lawson, up the slope of Columbia Mountain to reach their mines. March 1, 1884, is the date attributed to the camp's beginning. It was initially named Daileyville for James Dailey, manager of a nearby mine. There is an indication that it may have also been named Chinn City for no apparent reason. At any rate it was renamed Silver Creek (for the creek of the same name) shortly thereafter.

Of the several mining properties on Columbia Mountain and Saxon Mountain, most notable are the Reynolds properties, and those of the American Sisters Group (originally Seven Sisters).

Silver Creek was a small camp throughout its duration. Its school had about 25 pupils enrolled. In the early 1900s the mines dwindled and so did Silver Creek. The mining camp high above Clear Creek Valley became a ghost town.

FREELAND

Location: 5 miles west of Idaho Springs via U. S. Highway 40 and Trail Creek Road (County Road 136)

During the early days of mining activity in the Clear Creek Valley, a couple of good strikes were made along Trail Creek. One was the Freeland Mine which was discovered by John M. Dumont (see Dumont). The Freeland produced well immediately. A settlement eventually began near the mine in 1877. It was originally called Trail Creek Camp, then renamed Freeland.

In 1879 Dumont sold the Freeland Mine to John W. Mackay who had previously struck it rich on the famed Comstock Mine in Virginia City, Nevada. The mine had a high yield during his ownership — and he rewarded his manager well with a $25,000 per year salary, which was a great deal of money in the 1880s.

As the Freeland, Lone Tree and other mines prospered, so did the community. There were about eighty homes, two stores, a saloon, and a public school — but no church. On those occasions when a preacher could be coaxed into making the steep climb to Freeland, the townspeople found a place to hold a service. Additionally, there were several mining buildings. Freeland even had a "suburb" called Bonito which consisted of a group of mills and houses. A smelter was constructed in 1884. During 1887, when Freeland was at its peak, it had a population of 447.

Freeland remained populated well into the middle of the twentieth century. It's longevity can be attributed to successful mining and the mill which was constructed on Lamartine Tunnel property.

ALICE

Location: 10 miles northwest of Idaho Springs via I-70 and Fall River Road

Below St. Mary's Glacier on Kingston Peak and alongside Fall River, lies the town of Alice. Placer gold was discovered along the river in the early 1880s, and three camps sprang into being in the region — Alice, Fall River, and Silver City. Silver City, which was located on the upper Fall River, had a very short life. Fall River emerged near the junction of the river for which it was named and Clear Creek. Alice (originally known as Yorktown) was the largest of the camps. As more and more gold was found, a stamp mill was constructed and Alice became a substantial town.

There was much mining activity along Fall River. The Alice Mine was the area's best producer and operated off and on for many years. It was named for its owner's wife — as was the town. A sizeable glory hole lies near the townsite as a remembrance of the mining days.

Today, the centerpiece of Alice is its little white schoolhouse, built in 1915, which is surrounded by several cabins — some old, some not so old.

YANKEE

Location: 12 miles northwest of Idaho Springs via I-70 and Fall River Road

The shortest route from Central City to Georgetown was the road over Yankee Hill which passed the gold camp of Yankee, down through Alice, and on to Idaho Springs and Georgetown.

Yankee was a late-blooming camp. Claims speckled the hillsides. Many log cabins sprang up. A post office was established at Yankee in November 1893.

When one mining company acquired the rights to a great many properties, it usually raised eyebrows in mining circles. Such was the case in Yankee, and it boosted its growth. A mill was built to process ores from the various mines, including the Gold Anchor, the area's best producer.

After the mining dwindled a few people hung on, for the camp was still a traveler's stop. The post office was discontinued in February 1910.

NINETY FOUR

Location: 11 miles northwest of Idaho Springs via I-70 and Fall River Road

The Ninety Four Mine was located in 1894 — and named for the year of its discovery. The mining community which grew up around it followed suit and was simply named Ninety Four.

The camp is located on the west slope of Yankee Hill midway between Alice and Yankee. It is situated in a most picturesque spot — opposite St. Mary's Glacier across the canyon, and with a panorama of mountains including 14,000 foot Grays and Torreys peaks to the south.

The camp is strung out on a single road, and is little more than a ledge on the hillside. Its structures vary from neatly trimmed frame to conventional log.

Ninety Four was preceded by both Alice and Yankee. Because of their close proximity both were relied on for commerce.

Today, new homes cover the area around Silver Lake below Ninety Four. New construction has climbed the hill to the point where roof tops have virtually reached the mining camp. One day soon Ninety Four may be plowed under to make room for this generation or the next.

LAMARTINE

Location: 8 miles west of Idaho Springs via U. S. Highway 40 and Trail Creek Road (County Road 136)

Although there were many mines in the surrounding hills, for the most part Lamartine was a one mine camp. It all began with four partners named Cooper, Shavanne, Medill, and Bougher who while searching for the source of Trail Creek, discovered float gold and a rich fissure vein. They promptly staked a claim — and called it Lamartine. After his death, Bougher's widow sold her 25% interest to Peter Himrod, her brother-in-law. He paid $250 for the share. Furthermore, he purchased Cooper's share for $25, and Shavanne's for $5. Medill disappeared and forfeited his share. Himrod invested much capital into the mine. After his death, his son also invested heavily into the property. The ore however, continued to be sub-par and yielded little. Young Himrod sold the mine for $360. The new owners suddenly struck it rich.

Immediately thereafter the town of Lamartine sprang into existence. And it did so down the steep hillside adjacent to the mine. Lots were terraced and roof tops were at the same elevation as the footings above. The town began in 1887 and prospered through the '90s and early 1900s maintaining a modest population of a few hundred. The mine died and the town died. The mine bounced back and continued to flourish for years. The newer mining properties sit well below the old camp — which never bounced back.

The site of Lamartine is situated at an elevation of about 10,500 feet midway between Trail Creek and Ute Creek. It can be reached (with difficulty for those who have never been there) by climbing south from Trail Creek Road or by climbing north from Ute Creek Canyon. Today, the hillside meadow contains many empty terraced lots, others with foundations, and some with partial walls. The hike is eventful, and the scenery spectacular.

CLEAR CREEK COUNTY 47

Looking west across Idaho Springs in the 1870s. *Colorado Historical Society.*

Dumont was originally named Mill City.
Denver Public Library, Western History Department.

Empire in 1861. Volunteers of Company "G", 1st Colorado Calvalry, are drilling in the background. *Denver Public Library, Western History Department.*

North Empire in 1864. *Colorado Historical Society.*

CLEAR CREEK COUNTY 49

Waldorf was built by Edward John Wilcox.
Denver Public Library, Western History Department.

Georgetown was a log and frame community in the 1870s.
Colorado Historical Society.

The original Windsor Hotel of Mary Lampshire, in Silver Plume. The building was destroyed in the fire of 1884 and rebuilt as the New Windsor. *Denver Public Library, Western History Department.*

Mill at Lawson on the Colorado Central Railroad. *Colorado Historical Society.*

William Stephens' store and saloon at Freeland.
Denver Public Library, Western History Department.

Looking east across the townsite of Yankee.
Denver Public Library, Western History Department.

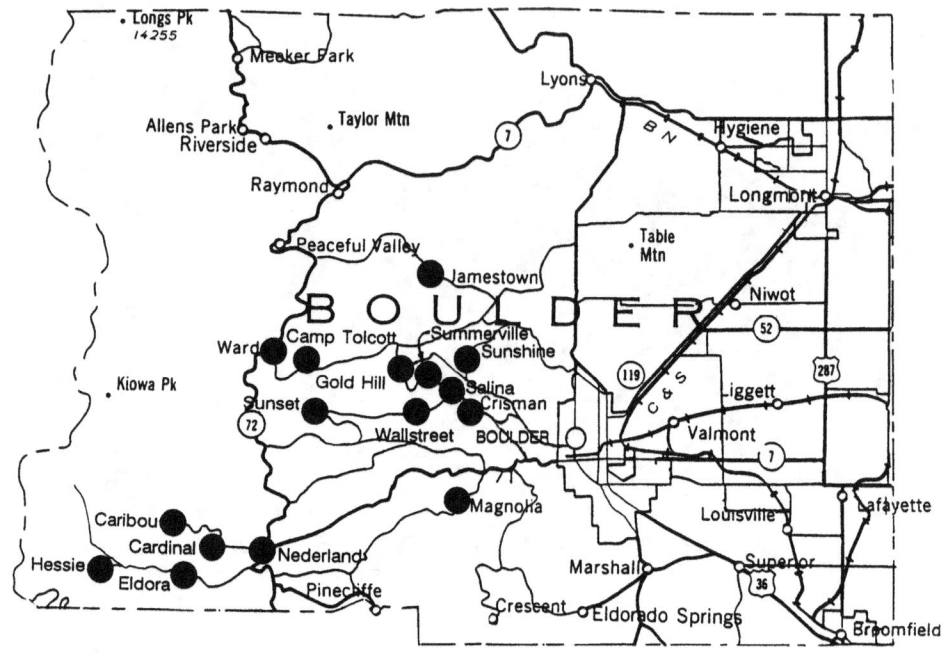

BOULDER COUNTY

The earliest settlers of Gold Hill established a law prohibiting the sale of liquor. Intoxicants flowed, however, following the rebirth of the community in the 1870s. This must have pleased Denver newspaperman Eugene Field (creator of Wynken Blynken and Nod) who was a hardy drinker and frequented Gold Hill's hotel. On one of his visits, after partaking of a little whiskey for "inspiration" Mr. Field penned his poem about Gold Hill entitled "Casey's Table d'Hote".

GOLD HILL
Location: 12 miles northwest of Boulder; 7 miles east of Ward

On October 27, 1858, a group of pioneer prospectors led by Captain Thomas Aikins established a winter camp alongside Boulder Creek at the mouth of Boulder Canyon and named the site Red Rock. An unusually mild winter allowed gold-seeking excursions from Red Rock into the adjacent mountains. One such party ventured through Sunshine Canyon, climbed to the ridge above Left Hand Creek, and set up camp at the site to be named Gold Hill on January 15, 1859. On the following day rich color was discovered in the streambed which was aptly named Gold Run. The news was carried back to Red Rock and from there spread rapidly.

Several hundred prospectors swarmed into the vicinity in search of the lodes which were the source of placer gold. Captain J. D. Scott located the first quartz vein, and the claim became known as Captain Scott's Discovery. David Horsfel, William Blore, and M. L. McCaslin found an important lode which they christened the Horsfel. Other important discoveries were the Twins and the Alamakee. The Alamakee was discovered by John Hitchings who considered it of little value and sold the mine for $25. The Alamakee yielded $125,000 to the purchaser in a very short time. The Gold Hill Mining District was soon established.

Initially there were two neighboring camps called Gold Hill and Gold Run. The first laws were recorded on July 23, 1859 whereby the camps of Gold Hill and Gold Run were united. Also on this date, the mining district was organized and the first mining laws were created. At the time, this area was part of Nebraska Territory, so the original laws were recorded as: "Gold Hill District Laws of 1859 — Mountain District No. 1, in Nebraska."

The components for a small quartz stamp mill were transported by teams of oxen up Left Hand Creek. The mill was built in the fall of 1859 and thought to be the first of its kind within the future state of Colorado. Later, a six-stamp mill was erected by the owners of the Horsfel.

A raging forest fire raced toward Gold Hill in the fall of 1860. Its advance was so rapid that residents only had time to grab a few personal effects and dive into prospect holes in order to save their lives. The fire scorched the west slope of Horsfel Mountain and destroyed much of the original settlement of Gold Hill.

By 1861 Gold Hill had become fairly deserted with only about 50 men left in the entire mining district. The placers had washed out and the rich surface ores had disappeared. Things would soon change, however. In 1872 Joseph Stepler discovered tellurium and staked a claim — the Red Cloud. A new rush was on. Another important discovery which followed was the Cold Spring. On the south side of Horsfel Mountain a fellow named Robinson located the Cash Mine. The Slide Mine, located on the north slope, ultimately had a greater yield than any other mine in the district. It also tunneled to a depth of 1,000 feet.

Whereas the original camp was located on top of Horsfel Mountain, the new Gold Hill began to materialize during the 1870s at its present location. One of the first buildings constructed in town was the Red Cloud Boarding House. A schoolhouse was built in the fall of 1873 with Hannah C. Spaulding as its first teacher. Within a short time there were three general stores, a grocery and meat market, drugstore, barber shop, doctor's office, and surveyor's office. There was no church, but religious services and Sunday School were held at the schoolhouse. In 1873 a thirty-room log hotel was constructed by a hotel proprietor named Wentworth. Today, one may step back in history by visiting the buildings of the Gold Hill Inn (the Bluebird Lodge as the hotel was renamed in 1921) and the Gold Hill Inn Restaurant located next door.

Peabody's Seltzer was exported far and wide from the seltzer house two miles east of Jamestown.

JAMESTOWN

Location: 12 miles northwest of Boulder via County Roads 75 and 94

Prospector and cattle rancher George Zweck built the first cabin in 1860 at the site which would later be Jamestown. Hundreds of fortune-seekers flocked into the vicinity following the discovery of rich galena veins in 1864. The camp, which blossomed around Zweck's cabin, was originally known as Elysian Park and then called Jimtown. When a post office was established in January 1867, it was officially called Jamestown contrary to the wishes of the community. Likewise, Jim Creek and Little Jim Creek (which meet at the townsite) were later renamed James and Little James creeks.

Shortly after the post office was established, the first boom was over and most of the people had gone. The Golden Age Mine was located in 1875 by Frank Smith and Indian Jack. They sold the property for $1,500, and later it became the best producer in the district. The Wano Mine (originally called the Bueno, for Bueno Mountain) was also discovered the same year. Mills were constructed for both mines. Some other mines in the vicinity were the Lily of the West, Argo, Humboldt, Burlington, Stanley, Red Spruce, Gold Cross, Sentinel, Earl, and the Cracker Jack. Across Porphyry Mountain was the Longfellow Mine.

With all the new mining activity Jamestown was busting at the seams. The boom of the 1880s far surpassed anything the community had previously seen.

Both the Evans House and Martin Hotel ran at capacity. For a miner's leisure time there were several dance halls, thirty-three saloons, and gambling dens. There was also an adequate number of parlorhouses and cribs, most of which were located in an area of town known as Bummerville.

Jamestown had several "suburbs." Balarat, north of Jamestown, was the largest and had some longevity. Camp Providence grew up at the John Jay Mine. Gresham and Camp Enterprise were also small population centers. Springdale, known for its mineral springs, had several good mines such as the King William, Grand Central, and the Copper Blush.

A devastating flood occurred in June 1894, and washed away much of Jamestown. The church, several businesses, and many homes were destroyed. The raging waters continued downstream to demolish Springdale as well.

Area mining has continued through much of the twentieth century. Today Jamestown is a quiet, peaceful community in a lovely setting.

WARD

Location: 14 miles north of Nederland on State Highway 72

When prospectors continually found good color along Left Hand Creek, they were certain that richer stuff existed in the surrounded hills. In 1860 Calvin M. Ward located a rich claim — the Miser's Dream — and the excitement began. The famed Columbia vein was discovered in 1861 by Cy Deardorff. The vein which yielded $5,000,000 was lined with rich claims. The Utica and Ni-Wot were the best properties. The Utica, another Deardorff discovery, produced gold worth $200 per ton, yielding an average profit of $10,000 per month. Some individual carloads were worth $5,000. The Ni-Wot claim was originally thought to have little value. It was purchased for $50 and sold two years later for $15 to W.A. Davidson and Samuel R. Breath. The mine paid them $50,000 in a few short months. The property continued to change hands but yielded an estimated $1,250,000 to the various owners during its day. Other claims along the lucrative vein were the East Columbia, Center Columbia, Baxter, Austin, and Idaho mines.

The sides of Left Hand Canyon (named for the Arapaho chief) were very steep. Before the railroad arrived it was very difficult for wagons to haul in supplies and equipment and also to ship out ore. The town of Ward is located in a basin. Some people say that the area is still so rich in minerals that the town should be razed and the "bowl" turned into a glory-hole. A post office was established in 1863 as Ward District. The town which grew to a population of about 1,000 was finally incorporated in 1896.

The Colorado & Northwestern Railway (originally spelled North-Western) arrived into Ward in June 1898. The rails twisted their way from Boulder westwardly to Left Hand Canyon, and was dubbed the "Whiplash Route." A "special" excursion arrived at Ward on June 28th amid much fanfare. Aboard were Governor Alva Adams and other dignitaries who had traveled to Ward for the celebration. An amusing sidelight to this occasion occurred when the train stopped at the Big Five Tunnel (the old Ni-Wot Mine) so the guests could experience a unique trip into a mine. A line of candles lighted the way. To enhance the thrill, miners detonated some shots in a branch tunnel. The blast of air from the shots blew out every candle and left the tunnel in total darkness. The thrill had been greater than anticipated — both to the apprehensive guests and the amused miners.

An energetic and hardworking 15 year old named Otto W. Carrow leased the Star Mine in 1888. After finding rich ore, Carrow went on a spree in Denver. He checked into the elegant Windsor Hotel, wined and dined, ordered a tailor-made suit topped with a derby hat, and purchased a bicycle. Aware of the newfound riches, and because of the boy's age, the property lessor had the lease declared invalid. Carrow, disappointed but satisfied, had netted $8,000 in a few short weeks.

From the time the first sawmill burned to the ground almost before it had produced any lumber, Ward has been plagued by fires. One conflagration destroyed the new Ni-Wot Mill in 1866. Other fires have occurred throughout the town's history, with the most disastrous being in 1900. The fire, which occurred at night, was presumed to have started from a can of hot ashes behind the McClancy Hotel and destroyed the entire business district (53 buildings) before it was brought under control. The *Ward Miner* stated: "Not a store, hotel, saloon, restaurant nor a business house of any sort escaped the flames." Townspeople saved the schoolhouse by keeping it drenched with water. Total property damage was $85,000, while insurance coverage was less than $7,500.

Ward was only partially rebuilt. A few landmarks remain. The old Congregational Church (later to become the Community Church) still stands like a centerpiece in the community. The old Catholic Church was converted into a tractor barn.

CARIBOU

Location: 4 miles west of Nederland

Colonel Sam P. Conger, a frontiersman and prospector, discovered silver ore at this location in the early 1860s. He didn't realize the value of what he had found, however, for several years. While in Laramie City, Wyoming, in 1869, he spotted a railroad car with similar ore being protected by armed guard. He was advised that the shipment was valuable silver ore from the Comstock Lode at Virginia City, Nevada.

Conger traveled back to Central City to recruit help. Accompanied by William Martin from Black Hawk, George Lytle, a Canadian, and three others, Conger returned to the site of his original discovery and staked the claim that became the rich Caribou Mine.

During the ensuing months the area became saturated with new claims, and the town of Caribou was platted. By 1871 there was a permanent population of nearly 1,000, which tripled during the short summer months.

Snow covered the ground nine months out of the year, and the town located at an altitude of 9,905 feet had a rough existence. Harsh winter winds and heavy snow-drifts made living difficult. Scarlet fever, diphtheria, and small-pox epidemics took a heavy toll on the community, which was also ravaged by two fires.

Caribou had two fine hotels, the Sherman House and the Planter's House, both of which were known for their hospitality and fine cuisine. Also included in the town's two-dozen business establishments, were several stores, saloons, pool halls, and brothels. In 1881, Caribou's "more-virtuous" citizens ran the "less-virtuous" ladies out of town in an effort to clean up the community. Many filtered back into town over a period of time. Caribou's population reached 3,000 at the height of the silver boom in 1875.

The Caribou Mine was the largest producer — yielding nearly $8,000,000. Other good mines were the Poorman, Sherman, 7-30, No Name, Idaho, and the Native Silver. Caribou's wealth was highlighted on April 28, 1873 when a path of Caribou's silver bricks was laid in front of the entrance to Central City's Teller House for the grand entrance of President Ulysses S. Grant. An exhibit of Caribou silver at the Philadelphia Centennial Celebration in 1876 brought additional attention to the town. Several pillars of finance and fortune were associated with Caribou mining — Horace Tabor, of Leadville fame, Jerome Chaffee, David Moffat, and R. G. Dun, founder of Dun and Bradstreet.

Production from the mines declined during the 1880s. The devaluation of silver in 1893 dealt the community a blow from which it never recovered. Two fires damaged much of Caribou. The first occurred in 1879. After the second fire in 1899, rebuilding was meager. The post office closed in 1917, but some mining continued on. The Hendricks Mining Company is still producing silver at Caribou. All that remains of the town, however, are a few stone walls and one lonely log cabin.

CARDINAL

Location: 3 miles west of Nederland

The Boulder County Mine was located in 1870 by Sam Conger (discoverer of the Caribou). A mining camp was established on the newly-built Coon Trail. Cardinal City was platted by J. D. Peregrine. Parallel streets were numbered. Perpendicular streets had names such as gold, silver, and quartz. There was a boardinghouse, post office, and an assay office, as well as several stores and saloons. The town became a wagon stop and stage station for transportation up and down the Coon Trail. Population, which was 200 in 1872, increased to about 1,500 during the peak years between 1878 and 1883.

When the city fathers of Caribou closed the brothels and told their "ladies of the night" to hit the road — they didn't go far. They moved two miles down the road and established a new "red-light district" at Cardinal City.

During the 1890s Cardinal City (often called Old Cardinal) faded rapidly. About one mile down Coon Trail, Cardinal (New Cardinal) was emerging to take its place.

After the turn of the century, during the tungsten boom, the railroad arrived at Cardinal. The community was sustained for a while by the Colorado & Northwestern Railway. Cardinal (New Cardinal) is the site today of several summer homes.

"A mail-order marriage is trickier than braidin' a mule's tail."

NEDERLAND

Location: 16 miles west of Boulder on State Highway 119

A camp was established at this location in 1869 and named Brownsville for its founder N. W. Brown. The camp was also briefly called Dayton. When the Breed and Cutter Silver Reduction Works was built on Middle Boulder Creek to serve the Caribou Mines, a post office was also established — as Middle Boulder (September 13, 1871).

Because of its low, flat location on the Boulder-Caribou Toll Road, Middle Boulder quickly grew as a stage station, shipping center, and supply town. In 1873 a Dutch company purchased the Caribou Mine, the mill, and other mining properties, which they operated as the Mining Company Nederland. In honor of these Dutchmen and their homeland, the town's name was changed from Middle Boulder to Nederland and officially incorporated on February 27, 1875.

When industry realized the value of tungsten for strengthening steel, Nederland experienced its greatest boom. Tungsten was everywhere — on Tungsten Mountain to the south, to the east at Tungsten Camp (Steven's Camp), and to the north at Lakewood which was the site of the largest Tungsten mill in the United States. Within a few years tungsten production from the Nederland area topped one million dollars annually.

During the boom, boardinghouses and hotels, like the Western, Antlers, Hetzer, Cory, and the Sherman House, rented beds in shifts. Additionally, miners slept wherever they could rent space for their bed rolls. Restaurants allowed customers twenty minutes to eat their meals.

Sixty percent of the United States' tungsten was produced from the Nederland area between 1900 and 1915. Shortly thereafter, the market for tungsten rapidly declined — and so did the town. Today, there is still a little mining activity in the area. Also, Nederland is striving to enhance its image as a tourist area.

SALINA
Location: 8 miles west of Boulder, or 1 mile northwest of Crisman via County Road 118

In 1873 a group led by O.P. Hamilton established a camp near the confluence of Gold Run and Fourmile Creek. He named the settlement Salina for his home in Kansas.

During the ensuing three years many mining claims were staked in the vicinity. The best mines were the Black Swan, Melvina, and Shamos O'Brien. Other producers were the Tambourine, Bessie Turner, Emancipation, Baron, and the Richmond. By 1875 the area boasted over 100 mines, which was nearly one mine per capita for the population of Salina was 112. Captain West, who owned the Shamos O'Brien Mine, constructed a reduction works.

Salina had a total of three mills, several stores and saloons, a toll station, church, schoolhouse, and a hotel — the Salina House. The Salina String Band provided entertainment for the community.

The post office which was established in 1874 was discontinued in 1925. Salina continues to prosper, however, as a community of predominantly summer residents.

SUNSHINE
Location: 8 miles west of Boulder via Sunshine Road

Prospectors searched the hills around the present site of Sunshine as early as 1859, but nothing much happened until good tellurium ore was discovered at nearby Gold Hill in 1872. During the following year, D. C. Patterson located a rich lode which he named Little Miami. Other strikes quickly followed such as the Sunshine, Grand View, White Crow, and the Inter Ocean Mines (which won acclaim at the 1904 World's Fair in St. Louis). With these discoveries the little town of Sunshine sprang to life.

George Jackson (also see Idaho Springs) and Hiram Fullen (also see Magnolia) discovered the American Mine which was to become (after they sold it) the biggest bonanza at Sunshine. Believing the mine had played out, Jackson and Fullen sold the property to New York hotel owner, Hiram Hitchcock, for the sum of $17,500 (approximately the same amount as the property had produced at the time of the sale). Almost immediately the mine began producing on a large scale for Hitchcock. Within two years it grossed nearly $200,000. Before the American played out, it yielded a total of $1,500,000. It made a wealthy man of Hiram Hitchcock and financed his New York hotel in the process.

Peter Turner was the first permanent settler at Sunshine. His daughter, who was the first child to be born there, was appropriately named Susie Sunshine Turner. The community grew quickly and within a short time there were a dry goods store, drugstore, blacksmith shop, two grocery stores, two saloons (which always closed on Sundays), and a newspaper — the *Sunshine Courier* (which began production May 1, 1875). There were also three hotels — the Grand View, the Howard House, and the Forest House. Room and board averaged about one dollar per night. Governor John L. Routt was guest speaker at an 1876 Republican gathering held in the Sunshine schoolhouse.

Some mining continued off and on for years, but only minimally compared to the glory days of Sunshine. The population which peaked at about 1,200 in 1876 decreased by one-third the following year — and, it continued to decline.

The Logan Mine near Crisman was the richest in Boulder County.

CRISMAN

Location: 7 miles west of Boulder on County Road 118

Free gold was found near Crisman in the year 1874. More good discoveries followed, and a city was platted close to the mines. Obed Crisman, an early settler, built an ore concentration mill. It is for him the town is named.

The Logan Mine, said to have some of the purest specimens found anywhere, and the Yellow Pine Mine were the area's top producers.

The Grand Republic Mine on nearby Fourmile Creek was owned by a Frenchman, Francois Ardourel. Being an aristocrat who loved fine wines, he built a large wine cellar which extended from under his house into the adjacent hill. Following the flood of 1894, the railroad tracks were rerouted to higher ground behind Ardourel's house, chopping right through his lavish wine cellar. The house and the "small" wine cellar which was left still stand in Crisman.

A. S. Coan, superintendent at the Logan Mine, managed the drilling of a 2,000 foot tunnel in 1908. After nothing was found, the owners ordered the digging to stop. Coan leased the property and promptly uncovered $1,500 in high grade ore. The lease agreement was quickly revoked. Did Coan know the ore was there all along?

Crisman never did grow very large. By May 1918 the last inhabitants had left and the post office was closed. As is the case with many "ghost towns," a revitalization occurred. Today Crisman is a peaceful little community nestled in the gorge beneath Arkansas Mountain.

Southeast of Sunset, near Sugarloaf, float gold was discovered in a potato patch in 1902 by a prospector named Niles. Over $20,000 was taken from the Potato Patch Mine.

SUNSET

Location: 14 miles west of Boulder, or 4 miles west of Wallstreet via County Road 118

Sunset was first established as a gold mining camp. The Poor Woman, Free Coinage (of which Gov. John L. Routt was part owner), and Scandia were all productive mines which were worked for several years.

The Columbine Hotel was the center of activity at Sunset. It was the meeting place for residents and visitors alike.

For awhile Sunset was the terminus of the railroad from Ward. After an extension was completed from Sunset to Eldora — known as the Switzerland Trail Line — the community became a popular stopover. The Columbine Hotel, it is said, had fifty people for every meal.

Above Sunset, on the road to Gold Hill, is a site where a ghost may be heard on clear, bright nights. According to legend, a young suitor built a white house — known as the Honeymoon House — for his fiance. Before they could be wed, however, she died. When the moon is full she can be heard whispering her love to her young man.

William La Shell, the Dalton & Sullivan Company, and Adolph Alpert each owned general merchandise stores. The meat market was operated by W.T. Linticum.

WALLSTREET

Location: 10 miles west of Boulder; 3 miles west of Crisman

Alongside County Road 118 in Wallstreet, stands an imposing stone tower, which looks much like something that should have been built on another continent, in another era. The structure was part of a mill constructed in 1902 by the Wall Street Gold Extraction Company. The structure was a storage bin for cooling pulverized ore, as part of the oxidation and extraction processes.

Mining was prevalent in the early 1890s, and a townsite began to grow. A post office was established October 31, 1895 in Delphi — the original name of Wallstreet.

In 1897, Charles W. Caryl arrived at the townsite with an ambitious scheme to purchase claims from local miners and then partially pay them back by giving them stock in his new Wall Street Gold Extraction Company. The town was renamed Wall Street (eventually Wallstreet). Regardless of the connotation of the name, the stock plan flopped. Poor management resulted in the mill's closure in 1905. Milling activity occurred off and on for many years, however.

Also of importance to the development of Wallstreet was the Nancy Gold Mine and Tunnel Company, which was organized in 1902 and produced a great amount of ore through its conglomerate of mines.

The population of Wallstreet peaked at about three hundred in 1903. The community has a few residents today.

MAGNOLIA
Location: 7 miles west of Boulder on County Road 132

While eating his lunch and tapping a rock simultaneously, Hiram Fullen discovered ore rich in tellurium. The find, which occurred in June, 1875, was named the Magnolia — as was the town that blossomed nearby. The settlement, which is located high above Boulder Creek, is scattered up and down a steep main street. Magnolia also had a small "suburb" called Jackson's Camp, which was named for George Jackson, Fullen's friend and partner in another endeavor (see Sunshine).

Besides the Magnolia, there were several other good producers — the Dunraven, Keystone, Mountain Lion, Lady Franklin, and Ben C. Lowell. Additionally there was the Queen Victoria, Atlantic, Young Magnolia, Jefferson, Fortune, Little Maud, and the Poorman. Closer to Jackson's Camp were the Wagner, Molly Gibson, Gold Farm, and other mines. Although there were many mines, they were surface mines and didn't yield for a long period of time. As a result, no mining syndicate ever moved in to control the whole area as they did in so many places. After the rich ore was gone, a cyanide mill was constructed to rework the low-grade ore.

The population of Magnolia peaked at about 300 then declined to 175 by 1887. Another Boulder County town, Gold Hill, had earlier passed a law forbidding saloons — and it seemed to work for awhile. Magnolia followed their example, and in 1876 said "no" to saloons by a vote of 173 to 27.

ELDORA
Location: 4 miles west of Nederland

The Happy Valley Placer was located in 1891 by Central City miners. A mining camp was established and named Happy Valley. The name was changed to Eldorado, and changed again to Eldora when the post office was established in 1897, in order to avoid confusion with Eldorado, California.

Gold tellurides were discovered on Spencer Mountain in 1897. When word got out that the tellurides resembled those found near Cripple Creek a few years earlier, the rush was on. The Enterprise, Terror, Bird's Nest, Clara, Bob Tail, Huron, and Revenge were a few of the mines which sprinkled the area. The Mogul Tunnel was drilled into Spencer Mountain below the mines to afford better access to the veins above. West of town, the Bailey Mill was constructed to process the area's ore. When the mill missed a pay day, irate workers found the manager and shot him.

The Colorado & Northwestern Railway (originally spelled North-Western) built an extension from Sunset to Eldora (known as the Switzerland Trail Line). The track ran through Pennsylvania Gulch, through Bluebird and Cardinal, and arrived in 1905 at it's terminus in Eldora. The railroad soon out-lived its usefulness and the tracks were removed in 1919.

During it's early years Eldora was a lively town. Gambling houses once ran through the night. The brothels were across the river, south of town.

No longer a lively town, Eldora today is a peaceful and picturesque community. The old Gold Miner Hotel still operates for tourists visiting the area.

HESSIE

Location: 6 miles west of Nederland; 2 miles west of Eldora

Hessie lies two miles west of Eldora where the two forks of Boulder Creek meet. Several of the original buildings remain in this community which lies in the valley below the road to the Fourth of July Mine.

Hessie, along with its sister camps Grand Island and Lost Lake Camp, was very dependent upon nearby Eldora. The three camps boomed during the 1890s. Several hundred people flocked into the area each year predominantly in the summer months.

Today, Hessie is privately owned. The camp is accessible, however.

SUMMERVILLE

Location: 10 miles west of Boulder via County Road 52

During the 1870s ore which averaged about $50 per ton simply did not cover a mine owner's overhead. This was principally due to the limited reduction processes in use at the time. As the turn of the century drew close, more sophisticated methods of treatment were in operation. Through the process of concentration, ores which were previously non-essential and less valuable could be removed by mechanical means. Ores averaging $20 per ton became profitable. Such was the case with the Black Cloud Mine and others near Summerville.

The camp at Summerville followed much the same pattern. While the mines yielded only low-grade ore during the '70s, miners lived in shacks and tents. A more substantial community cropped up in the late 1890s with the advent of more efficient milling methods and new discoveries. The settlement, which never had a business district, relied on Gold Hill for its commerce.

In addition to the Black Cloud Mine, others were the Cash, U.S. Bank, Hoosier Ledge, C.Q., and the Victoria. Charles Davis, owner of the Victoria Mine, was lying on his death-bed when his lessee discovered a new rich vein. Just a month later widow Davis sold the property for $125,000.

Much of Summerville, which steps up the side of Gold Run Gulch, looks as it did in the early 1900s.

CAMP TOLCOTT
Location: 6 miles east of Ward

In Left Hand Canyon near Hanging Rock, a strike was made by Colonel Wesley Brainard. The discovery which he called the Brainard Mine yielded high-grade ore. To assure the success of his endeavor Colonel Brainard solicited the assistance of Lyman J. Gage, a Chicago financier. Gage, Secretary of the Treasury during the William McKinley administration, and his group invested $700,000 into the Brainard Mine, the Giles Mine, and the community of Camp Tolcott.

Camp Tolcott was once generally referred to as Brainard Camp. An old J. B. Sturtevant photograph shows Camp Talcott spelled with an a, as do some other references and writings. The earliest histories spell Tolcott with an o.

Brainard, who also founded the nearby community of Quigleyville, assumed the position of superintendent of the Gage properties. The success of Gage and Brainard was due in large part to electricity. When the Giles opened in 1899, electrically powered steam drills were used. They were used for additional drilling at the Brainard Mine as well. Ores were carried from the mines to the mill by electric tramway. Power poles were installed and the community was illuminated by arc lighting.

Doc Vaughn Saloon at Gold Hill. *Colorado Historical Society.*

The McClancy Hotel was once a center of activity at Ward. *Colorado Historical Society.*

Caribou at the height of its prosperity. Pete's Saloon is at left.
Colorado Historical Society.

Between Nederland and Caribou lies the site of Cardinal.
Denver Public Library, Western History Department.

BOULDER COUNTY 67

Nederland at the turn of the century. *Archives, University of Colorado at Boulder.*

Customers of the Salina Inn, a saloon, pose with their drinks.
Denver Public Library, Western History Department.

The Crisman Mercantile Co., as photographed by J. B. Sturtevant. *Colorado Historical Society.*

Railroad depot at Sunset. *Denver Public Library, Western History Department.*

Guests and others pose by the boardinghouse at Wallstreet.
Denver Public Library, Western History Department.

Eldora's first post office was located in the home of Mrs. Lois Holzhauser.
Colorado Historical Society.

70 NORTH CENTRAL REGION

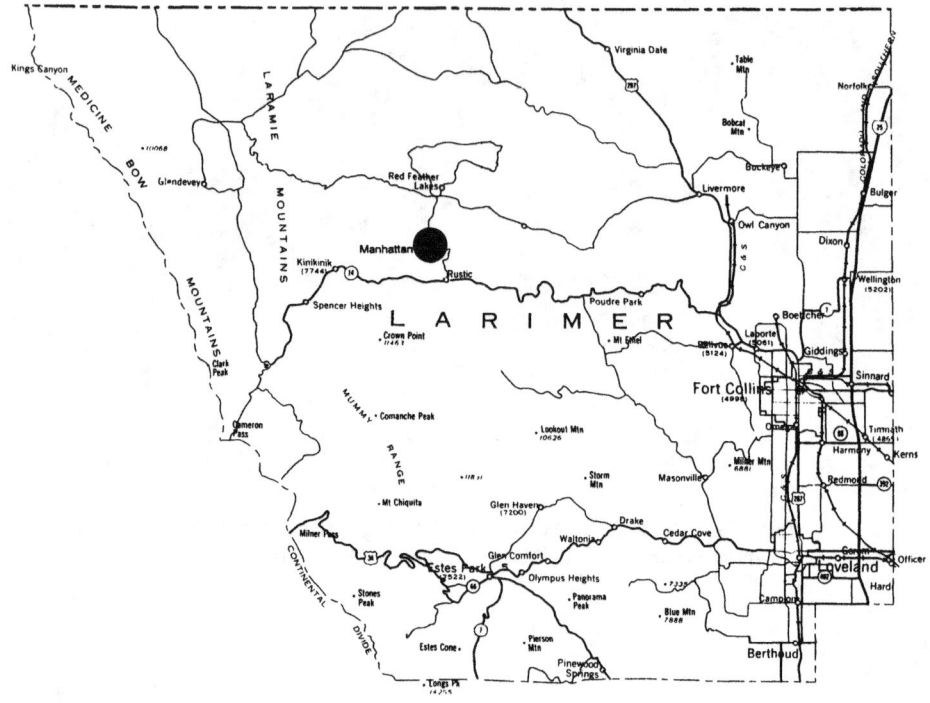

LARIMER COUNTY

"Broke is what you get when you let your yearnin's get ahead of your earnin's."

MANHATTAN

Location: Forty miles west of Collins on State Highway 14 - then 4 miles north

A short distance east of the future site of Manhattan, gold was discovered in 1885. The small camp Elkhorn (also called Elktown), with a population of about 100, sprang up around the Elkhorn Mine. As prospectors flocked into the area most of the activity shifted to the west. Several claims were established as was the town of Manhattan. It was a community that had great expectations — but a brief life.

About 35 log and frame structures were constructed down both sides of Main Street, including a saloon, stores, and a hotel. A post office was established in March 1887. Manhattan became the center of activity in the area.

Between Manhattan, Elkhorn, and Rustic Camp (to the south), the Elkhorn School was built. It served each of the aforementioned communities.

Although there was some good ore, most of it was subpar. Sizeable expenditures were made on many mines, such as the Emily, Monte Cristo, Little Tipsy, and Katy's Pet — to little avail. Because of the rather remote location of the area, shipping the ore out was costly. Many miners moved on.

Manhattan wasn't quite ready to die, however. In 1896 a rich vein 18 inches thick was discovered near the Cache La Poudre River. The community sprang back to life — with an estimated 300 residents. The new boom didn't last long either. The good ore played out quickly, and once again people packed up and moved. The post office was discontinued in December 1900.

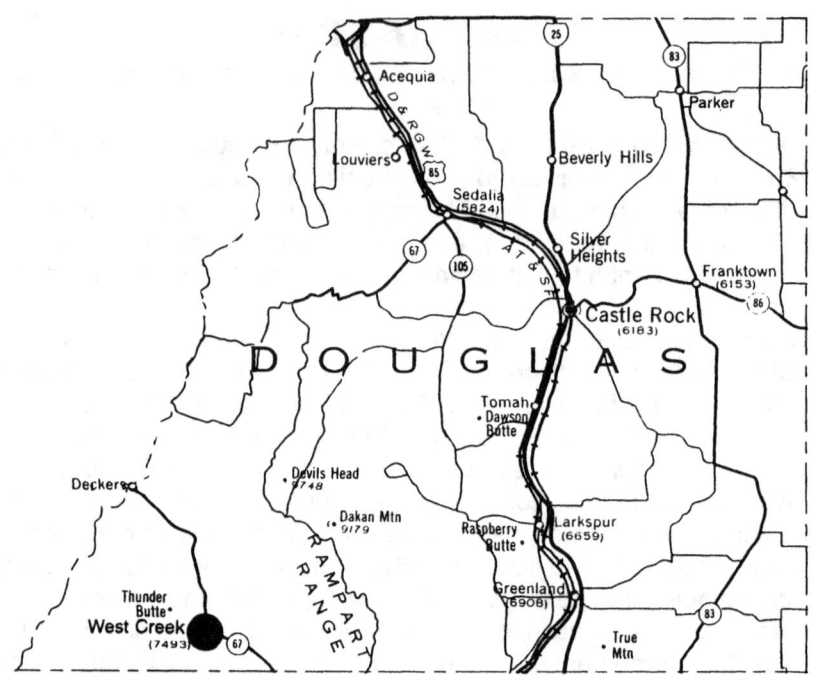

DOUGLAS COUNTY

"Prospectin' is a game of lost and found ... more lost than found!"

WEST CREEK

Location: *14 miles north of Woodland Park on State Highway 67*

Ranchers discovered a little gold in 1895 in the vicinity of West Creek and Trout Creek. Visualizing a bonanza similar to that which was presently occurring at Cripple Creek, prospectors rushed into the area. Several settlements sprang up in the vicinity including Pemberton, the largest of the camps, which was later to be renamed West Creek.

As most of the ore was low-grade, miners held a meeting to discuss development of the many claims. According to the *Silverton Standard* (Dec. 28, 1895): "Each man is to work so many days for nothing, with the understanding that he has no claim against the property worked upon, unless paying mineral is found, (in which case) his claim shall be a lien on the mine. The idea is to demonstrate what the camp is, by shipping ore as soon as possible, and all hands will turn in and aid in a movement which is for the benefit of all."

The Miners' Protective Association planned another project for the community. The goal was to drive a 500 foot tunnel into the most promising vein. Workers would earn shares of stock based on the number of shifts they worked.

There were two discoveries in May, 1896. A strike was made by Nick Miller in the Last Chance Mine near Pemberton. Another was made nearby at the Given's Ranch. The new finds brought additional fortune seekers to the area.

The town blossomed. Soon both sides of the main street were lined with buildings — about forty in all, including ten saloons. There was also telephone and telegraph service. Two Pemberton women were making a fortune baking bread and selling loaves at fifteen cents each.

Pemberton became West Creek (sometimes spelled as one word Westcreek). It should not be confused with another small camp of the same name which was one of the original camps in the vicinity and located nearby. It was insignificant in comparison, but confusing.

Area mining never met expectations and West Creek eventually declined. Today, the community is a mixture of newer cabins and the few remaining old ones.

West Creek as it once looked. *Collection of Dave Southworth.*

West Creek, which once boasted ten saloons, was originally named Pemberton. *Collection of Dave Southworth.*

CENTRAL REGION

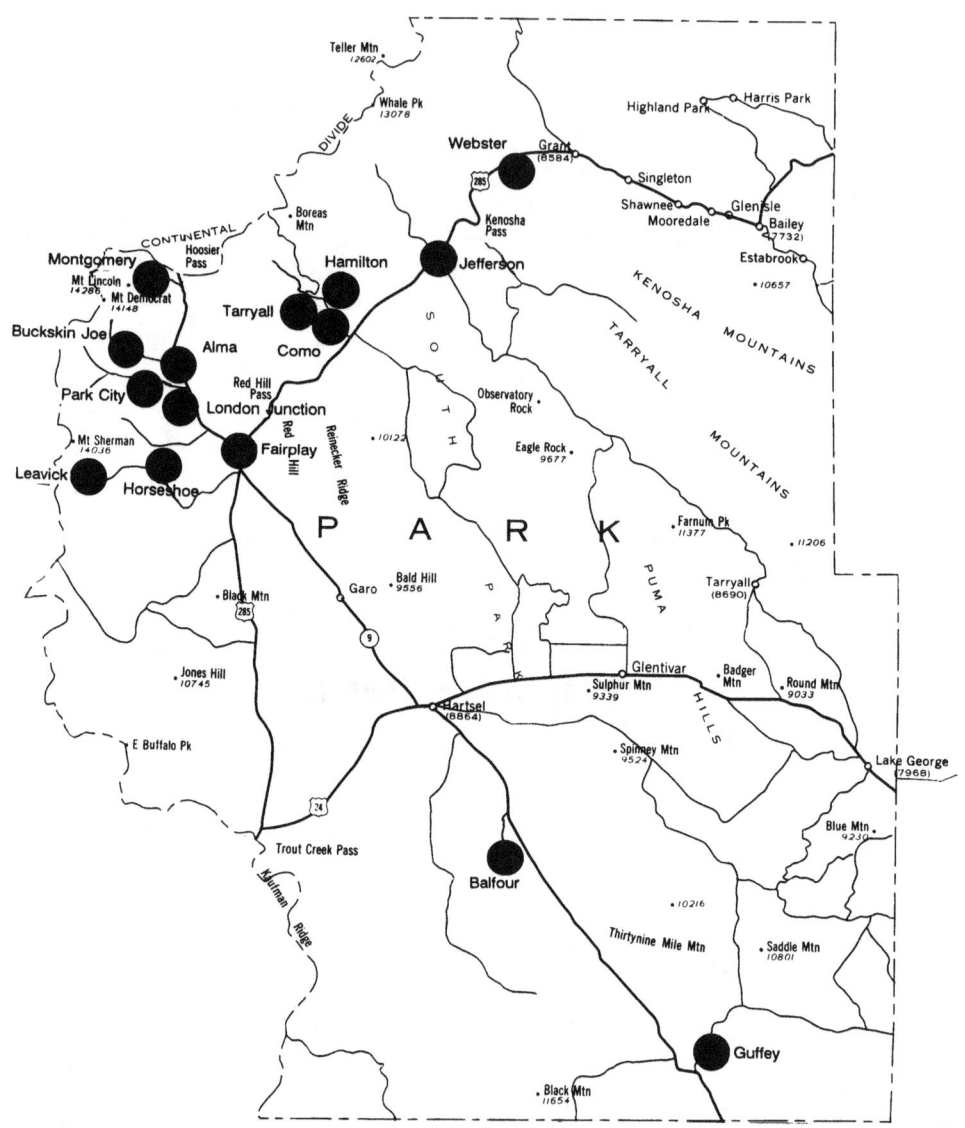

PARK COUNTY

"Trying to get even is a sure sign that someone's wrong."

FAIRPLAY

Location: 22 miles south of Breckenridge on State Highway 9

According to William B. Webster, his grandfather William Wilcox Webster named the town of Fairplay. One day while traveling toward the future town site, the elder Webster met a stranger on the trail. Webster noticed a bullet hole through the stranger's hat, but it was not mentioned by either party as they spoke briefly and then rode on. When Webster arrived at the tent colony, which was the future town site, he was informed of a duel that occurred several hours earlier. Evidently a stranger had entered a tent where a card game was in progress and accused one of the players of murdering his sister. He challenged the card player to a gun fight. Both men drew — the card players bullet struck the stranger's hat, while the stranger's bullet killed the card player. Realizing that this was the same stranger that he had encountered on the trail, he asked if the camp had a name. Advised that it did not, Webster said, "Let's call it Fair Play!"

Another story says that the miners were simply not going to have all the land-hogging, claim-jumping, and bickering typical of many camps such as Tarryall, which the miners called "Graball." They insisted on being fair and square — hence the name Fair Play.

Whatever the source of the name, it was originally two words. The town sprang up along the South Platte River when gold was discovered at the site in 1859. News spread fast, and prospectors came from near and far. Log structures quickly replaced tents as the town took on an air of permanence. By the time the placer mining played out, Fair Play had already become a substantial city. The town which evolved as a supply center for the vicinity, is located near the geographic center of Colorado. In 1867 it became the county seat of Park County.

The fateful "Tuesday Murder" occurred April 3, 1879. The Bergh House (now the Fairplay Hotel) hired a fellow named Thomas M. Bennett to work on a drainage ditch that ran down the street in front of the hotel. John J. Hoover was upset over the progress being made on the ditch which also passed in front of his billiard parlor. He had a drink or two then walked into the hotel lobby, found Bennett, uttered a few ornery words and shot him dead. Hoover was jailed, tried, and sentenced to eight years in prison by Judge Thomas Bowen. Enraged by the light sentence, the citizens took the law into their own hands and hanged Hoover from the second story window of the Court House until he was dead.

Soon thereafter, while awaiting trial for murder, a skinny lad named Cicero Simms boasted he was too light to be hanged. He was found guilty, received a public hanging, and was pronounced dead. Simms' two brothers had been locked up in jail during this ordeal to avoid trouble. They were released and took custody of their brother's body. That evening a stagecoach driver was stopped by two men who had an injured companion. The driver, who was forced to take the three to Leadville at gun point, later identified the injured man as Cicero Simms.

The city was wild and had a reputation for being so. The population is estimated to have been as high as 8,000 at its peak. In 1873 a fire destroyed much of the downtown business district. In 1924 the U. S. Post Office officially changed the name of Fair Play to Fairplay (one word).

On Front Street stands a monument to the burro Prunes. He was such a friend of the community and faithful companion to prospector Rupert Sherwood that at Sherwood's request he and Prunes were buried together. The inscription reads: "Prunes, a burro; 1867-1930; Fairplay, Alma, all mines in this district."

There are many interesting historical structures in Fairplay, including the aforementioned Court House (now the Park County Library), the Fairplay Hotel, the South Park Community Church (originally the Presbyterian Church) which was dedicated in 1874 by the famous missionary Sheldon Jackson, and several others.

Fairplay is the site of a marvelous restoration — the South Park City Museum. Thirty buildings were moved from the towns of Alma, Garo, Leavick, Buffalo Springs, Dudley, and other places. They were intermingled with five structures at their original locations and several others from around Fairplay. All are restored and full of thousands of authentic artifacts and furnishings. The South Park City Museum is very representative of an 1880s Colorado mining town.

"The three most fatal diseases in the west are smallpox, cholera, and the ignorance to argue with a mean-lookin', whiskey-drinkin', liar."

BUCKSKIN JOE
Location: 2 miles west of Alma

"Buckskin Joe" was the nickname for Joseph Higgenbottom, part-time prospector and part-time trapper. Higgenbottom, who always wore leather clothes, made a placer strike at this location in 1860. A year later he traded his interest in the camp for a horse, a gun, and satisfaction of his bar bill.

The *Rocky Mountain News* once described Buckskin Joe as South Park's "liveliest little burg." Half the residents are said to have made their living from the several saloons, gambling halls, and billiard parlors. Buckskin Joe (renamed Lauret, or Laurette throughout part of the '60s) had its own newspaper, stage office, theater — Laurette Hall, and the Bank of Stansell, Bond, and Harris. The popular Pacific House was one of four hotels. The town was county seat for about seven years before the seat was moved to Fairplay.

The Phillips Lode (named for M. Phillips, one of the earliest settlers here) produced over $300,000 in a two-year period. The discovery (made in 1861) was the best lode in Park County at the time.

The story of J. B. Stansell is an interesting one. He left Oro City, crossed the mountains, and became one of the first settlers of Buckskin Joe. Stansell, an original partner in the Phillips Mine, became immensely rich after M. Phillips sold him his share of the mine soon after its discovery. In fact, he was the richest man in Buckskin Joe only a year after he left Oro City.

H. A. W. and Augusta Tabor (of Leadville fame) lived in Buckskin Joe during part of the '60s. Together they operated a store while he served as postmaster.

Father John L. Dyer, an itinerant preacher, wandered into Buckskin Joe in 1861, with little more than a prayer book to his name. In this predominantly God-forsaken country, he spread the gospel wherever he could find a few men gathered together — whether in a gambling hall or a mine. Dyer was a dedicated missionary who covered most of the mining camps in the area afoot. His collections were meager, so he worked odd jobs to help pay expenses. He carried mail, hauled bags of gold dust, and even drove a stage for a while so he could continue to spread the word of God.

According to a legend (which seems to have no validity), there was a beautiful dance hall girl named Silverheels — so named because of the silver heels on her dancing shoes. During the cold winter of 1861, an epidemic of smallpox swept through the mining camp. Most of the miners became ill, and many died. Throughout the ordeal, Silverheels went from cabin to cabin caring for the sick and comforting the dying. She scrubbed and cooked and nursed. Finally she too was stricken by the illness. When the epidemic subsided, miners collected $5,000 in cash as a gift for Silverheels to show their gratitude for what she had done. The men carried the reward to her cabin only to find that she had disappeared. Years later a heavily veiled woman was seen in the cemetery weeping over the graves. When approached she quickly departed. Had Silverheels come back to mourn her friends? The miners certainly thought so. She had sacrificed her beauty and thereafter shunned her former admirers. A majestic peak, Mount Silverheels, was named in her honor. Although there may be no truth to the story, it clings to the history of Buckskin Joe.

Little remains of Buckskin Joe other than its cemetery, an arastra (a hollow stone basin used to grind ore), and a few mining ruins.

MONTGOMERY

Location: 11 miles north of Fairplay via State Highway 9

Not far from the trail that Thomas Jefferson Farnham blazed across Hoosier Pass 10 years earlier, gold was discovered in late 1859. As was the case with most of the earliest strikes in Colorado, people flocked to the area from great distances. The town of Montgomery sprang up the following spring. Before long there were two hotels, the usual stores and saloons, and a concert hall which was used for dances, plays, and other performances. Within a short time Montgomery was the largest town in the area.

Water-powered stamp mills and sawmills were constructed adjacent to the South Platte River on the west edge of town. More and more strikes were made. The Pendleton, Magnolia and the Montgomery were among the best mines.

Many of the rich finds were made on Mount Lincoln, named for our 16th President who was newly elected in 1861. It is said that Montgomery was one of Schuyler Colfax's favorite places. It is also said that Colfax, who was a member of Lincoln's cabinet, proposed to his bride-to-be at the summit.

A story is told about a prospector who erected a tent for his wife and children on September, then set out with another man to purchase supplies in Denver. The fought a blizzard on their return into South Park a month later. When the prospector reached his family he found them frozen to death.

The town, which peaked in 1861, declined almost as rapidly as it grew and was nearly a ghost by 1868. A short revival occurred when Leadville experienced its silver boom, but it too was short-lived.

Today, the remains of Montgomery are drowned in the waters of the reservoir which bears the same name.

PARK CITY

Location: 3 miles west of Alma in Mosquito Gulch

On Mosquito Creek by the entrance to the Gulch with the same name, Park City sprang into existence in 1861. It remained a small camp, however, for many years. The settlement which had nearly died before the road opened over Mosquito Pass began to grow in 1879. For a few years before the railroads were built, Mosquito Pass was an important route to and from Leadville. Although a mining town, Park City also became an important stage stop.

By the summer of 1880, fifteen substantial buildings had been constructed including the Park Hotel, a saloon, mercantile store, and boardinghouse. The community had a population of over 200.

When the railroad arrived into the area in 1882, the Mosquito Pass Road began to outlive it's usefulness. The stageline across Mosquito Pass was soon discontinued. Park City remained inhabited for many years, however, because of the mining activity in the area. Successful mines in the vicinity were the Orphan Boy, Hock-Hocking, and Brownlow.

One of the hotels was converted into a very popular brothel, and it's "girls" served miners from many of the surrounding communities.

Park City was sometimes called Mosquito. One prominent historian places the site of Mosquito four miles further west at the location of the North London Mill. Although there were a few cabins around the mill, it was not the site of Mosquito. Also, there was a halfway house built in 1879 on Mosquito Pass Road near timberline, which was a stagecoach stop and saloon where a few "lewd ladies" awaited weary travelers. It too was often referred to as Mosquito. The gulch, creek, road, peak, pass and mining district were all named Mosquito so it's confusion is understandable. The halfway house and an assay office from the North London Mine are two of the buildings which were moved to the South Park City Museum in Fairplay.

"Every jackass thinks he's got horse sense."

TARRYALL
Location: 2 miles north of Como on County Road 33

Gold as "big as watermelon seeds" was discovered along Tarryall Creek in July 1859 — or was it rediscovered? Prospectors arrived in the area and found a few abandoned log cabins. Could this have been the site where pathfinder John Fremont describes a gold discovery in 1849 by "Parson" Bill Williams? Even earlier in 1805, James Purcell made the first recorded gold discovery in the west near this location. The find is documented in the journals of Zebulon Pike.

A camp was established around the larger than normal claims. The several hundred inhabitants quickly gained a reputation of being inhospitable and of hogging all the best sites. Those that arrived later left in disgust, dubbing the camp "Graball."

The city was platted in 1861, and for a very short period was the county seat of Park County. They say everybody in Tarryall was a whiskey drinker — it's probably because there were many more saloons than normal for a town its size.

A private mint was established by John Parsons in the sixties which stamped out $5.00 and $2.50 gold pieces.

During its active years, the "Tarryall Diggings" produced approximately $2,000,000 in gold. Eventually gold became scarce and the miners moved on. Tarryall was totally deserted by 1875, a few years before nearby Como sprang into existence. Dredging operations along the creek prior to World War II buried the remains of the town under tons of rock.

"Jus' 'cause a panner is a mite whiffy don't mean he's 'fraid of water. Sometimes he uses it as a chaser."

HAMILTON

Location: 2 miles north of Como where County Road 33 crosses Tarryall Creek

Hamilton is another of the gold camps along Tarryall Creek which lies beneath the rubble of dredging operations.

It was founded, some say, by disgruntled prospectors who moved further downstream when there were no decent sites left at Tarryall. Unlike its "sister" camp, the welcome mat was out and Hamilton grew faster and larger than Tarryall.

The community had an element of "culture." It built a theatre which booked many plays and acts. A newspaper, *The Miner's Record,* was published in Denver and circulated in Hamilton.

When the supply of gold fizzled out — so did Hamilton. By the mid-'70s it was a ghost town.

LONDON JUNCTION

Location: 2 miles south of Alma via State Highway 9

London Junction and Alma Station (Alma Junction) grew at the intersection of the Mosquito Pass Road and the road from Fairplay to Hoosier Pass. It was a stop for travelers to and from Leadville to the west and Fairplay to the south. Five hundred yards separated the two communities which were virtually one.

After the railroad arrived a spur was constructed from London Junction through the little town of Park City, along the Mosquito Creek, and on to the London Mountain mining area. The North London Mill was the terminus of the London, South Park and Leadville Railroad which was completed in 1882. The North London Mine which was discovered in 1875, had its own boardinghouses and an aerial tram nearly two-thirds of a mile long. It produced over $8,000,000 in gold, silver, and other minerals.

London Mines Incorporated, owner of both the North and South London mines, built a concentration works at London Junction in 1883. Ore was transported by rail from London Mountain to the mill where it was processed.

Placer mining occurred throughout the vicinity and many miners lived in London Junction and Alma Station. There are a few residents in the area today as well.

It is believed that "The Bloody Espinosas" who terrorized parts of Colorado during the mid-1860s killed 6 of their 32 victims near the future site of Alma.

ALMA

Location: 6 miles northwest of Fairplay on State Highway 9

Alma originated nearly two miles north of the location of Alma Station (see London Junction) when a better route to Buckskin Joe developed from the Fairplay-Montgomery Road. At first it was no more than a junction house and traveler's stop. When the mines at Buckskin Joe played out some of the residents moved further down into the valley, and the community of Alma was born in 1872.

A post office was established in 1873, the same year it was discontinued at Buckskin Joe. Alma experienced some growth in the late '70s when prospectors came clamoring from Leadville over Mosquito Pass in search of new silver strikes. With the burst of new mining activity in the vicinity, Alma's location in the valley made it a desirable place to live. The community, whose population never reached 1000, had hotels and boardinghouses, a schoolhouse, newspaper, stores and saloons. There was also a smelter — a division of the Boston and Colorado Smelting Works.

Alma was a common name during the nineteenth century. Some people say the town was named for the wife of Abner Graves, a mine owner. Others say it was named for the daughter of the town's first merchant, a fellow named Janes. There are those who believe it was named for the wife of a town leader, or for the first baby girl born in the community. Maybe there were so many Alma's that the town was named for all of them.

Later in Alma's existence, a fire in 1937 started in one of the saloons. Before it was brought under control it had destroyed nearly a dozen buildings, including a garage which housed most of the town's motor cars.

Since its beginning, Alma has always been inhabited.

JEFFERSON

Location: 16 miles northeast of Fairplay on U.S. Highway 285

The earliest known photograph of Jefferson was taken while it was still a tent colony. The Denver, South Park & Pacific Railroad, which arrived into the area in 1879, can be seen in the background. Although no substantial construction took place at Jefferson prior to the arrival of the railroad, the community of tents actually arrived first.

History indicates that there may have been a "settlement" at this location as early as the mid 1860s, when prospectors crossed Georgia Pass after rich strikes were made in Georgia Gulch on the other side of the divide. There was a marked degree of activity along Michigan Creek and Jefferson Creek. Gold was found, but it didn't amount to much.

The camp emerged at the junction of the road from Denver to Fairplay, with the road to Georgia Pass. It is a matter of conjecture as to whether or not the settlement existed continuously until the late '70s.

The town which lies in the rather barren high mountain valley south of Kenosha Pass (and the creek) were named for Thomas Jefferson, our third President. The Jefferson depot has been renovated, and may still be "for sale."

Highwayman Jim Reynolds and his gang — raising money for the Confederate Army — robbed one of many stagecoaches near the future site of Como.

COMO

Location: 10 miles northeast of Fairplay

For several years prospectors traveled through the vicinity, but there wasn't much activity. A little gold mining was followed by some coal mining.

Prior to the arrival of the Denver, South Park & Pacific Railroad on June 21, 1879, the site was called Stubbs Ranch and was an important stage station. A depot was constructed, and a town was platted by George W. Lechner. The new town was called Como for nearby Lake Como which had been named by Italian coal miners after it's namesake in Italy. It was an ideal location to service the towns in South Park and connect with the stage line over Boreas Pass to Breckenridge. The depot was followed by many other structures.

In July 1879 the *Fairplay Flume* stated, "Thirty days ago there was no Como ... Between 60 and 75 tents and a dozen or more wooden buildings, for accommodations and businesses, are standing."

The advent of the railroad increased coal production in the area, and Como became not only a coal mining community but also a railroad town — and a wild one at that. In addition to its hotel (the Pacific), restaurants, and stores, Como had many busy saloons and brothels. The town was rowdy, with fights, racial bigotry and violence commonplace. On one occasion, a supervisor who hired some Chinese miners, was beaten up and run off by a group of Italians. They also ran the Chinese off. In 1893 many deaths occurred when an explosion devastated the King Cole Mine.

The most interesting sight for tourists visiting Como today is the roundhouse. The structure was built in 1881 and had subsequent additions and modifications. After the railroad was discontinued, the roundhouse had many tenants including cattle and horses — for it was used for a while as a stable. Como Roundhouse Preservation, Inc. is attempting to raise funds for restoration of the structure.

Several fires plagued Como over the years. In 1896 the Pacific Hotel burned down. The Como House (a new hotel) was built the following year to replace it. Most of the railroad shops burned down in 1909. A wooden addition to the stone

roundhouse burned to the ground in 1935, destroying two engines in the process.
The community remains inhabited today, but without the activity of yesteryear.

"If the coffee tastes like mud — just remember it was ground this morning."

HORSESHOE
Location: 8 miles west of Fairplay via U.S. Highway 285 and Fourmile Road

Horseshoe, which was also called East Leadville or Doran at different times during its existence, was established along Fourmile Creek in 1879. The first silver strike in Horseshoe Gulch was made at this location and was the catalyst for the growth of the community which incorporated in 1881. Its name is derived from the natural glacial cirque of Horseshoe Mountain.

Following the earliest discoveries the South Park Smelting and Reduction Works built a processor in 1879. Forty employees processed up to ten tons of ore per day. A second smelter (the McFerran) burned in 1881. Several top producing mines were located in the Horseshoe-Leavick vicinity including the Hill Top, Dauntless, Last Chance, Mudsill, and Crusader.

Estimates of the population of Horseshoe vary from 300 year-round residents to upwards of 800 during the summers. Two hotels, several saloons, and other business establishments sprang up. Additionally, two sawmills were constructed.

Although it realized a couple of revivals, Horseshoe only flourished about fifteen years. The devaluation of silver in 1893 spelled doom for the camp. A few residents hung on, but for the most part Horseshoe was well on its way toward becoming a ghost town.

When ores at the Mudsill Mine played out (so the tale is told) the owners salted the tunnel with silverdust then sold the property to the Lord Mayor of London for $190,000.

LEAVICK
Location: 12 miles west of Fairplay via U.S. Highway 285 and Fourmile Road

A short distance up the gulch from the town of Horseshoe, a mill was constructed below the Last Chance Mine in 1885, and adjacent to it a settlement began to take shape.

The camp, which had no name for awhile, was finally named in 1896 for Felix Leavick, a prospector who had purchased the Hill Top Mine four years earlier. A couple of suburbs also sprang up nearby. Mudsill was a small camp located about 3,000 feet away around the Mudsill Mine. A cluster of cabins on the rise above Leavick was dubbed New Leavick.

The Hill Top Mine was one of the best producers in the county. A long aerial tram dropped from the mine to carry ores to the Hill Top Mill which was built in the 1890s. The mill was also the terminus of the Denver and South Park Railroad narrow-gauge spur to Leavick, which was officially known as the Denver, South Park and Hill Top Railway.

The community had a post office, two saloons, a parlor house, a barber shop, a small schoolhouse, and a couple of stores. For a short while it also had a baseball team. Leavick had several hundred residents at its peak.

The Hoffman Brothers Blacksmith Shop and the Homestead House are two buildings which were moved from the town of Leavick to the South Park City Museum in Fairplay.

An easterner once said, "Immigrants heading west to look for gold were born silly and had a relapse."

BALFOUR
Location: 10 miles south of Hartsel via State Highway 9

About the same time as the Cripple Creek boom, gold was discovered at Balfour. Overly optimistic prospectors, with dollar signs in their eyes, imagined the area to be full of riches. Word spread rapidly, and hundreds of people arrived in a hurry.

A sawmill was constructed and buildings sprang up everywhere. There were three hotels — the Balfour, the Crawford, and the Clarendon, several stores and saloons, and a newspaper — the *Balfour News Weekly*. Daily stagecoaches arrived at Balfour from both Wilkerson Pass and Currant Creek Pass.

Many mining camps, during this period, were troubled with ethnic disputes. Balfour solved the problem by prohibiting Chinese and Italians from entering the community. Whether or not there was any other basis for this act of discrimination we do not know, but Balfour never had labor problems.

After four years of digging, with little to show for it, the community rapidly declined. The newspaper stopped publishing in 1897. By the turn of the century, Balfour was gone.

GUFFEY
Location: 30 miles north of Canon City via U.S. Highway 50 and State Highway 9

The community of Freshwater, as Guffey was originally named, began in the late 1890s during the height of the Cripple Creek labor wars. To avoid the dissention and violence created by the dispute between union and non-union factions, many miners packed up and moved further west in search of new gold strikes.

When prospectors located traces of gold, the Freshwater Mining District was established, and gold-seekers envisioned a new boom similar to that which occurred at Cripple Creek.

Several businesses were quickly established. Two hotels were constructed, a dance hall, blacksmith shop, livery stable, a general store, and five saloons. Freshwater also boasted something quite uncommon in the old west — a lady barber.

The community which grew adjacent to Freshwater Creek had been given a rather non-appealing name. When a senator from Pennsylvania offered the town $500 to change its name to his, the offer was accepted and the community became Guffey. When the small quantities of surface gold played out, prospectors packed up and moved on to find more promising areas. Four of Guffey's five saloons closed, and the town's population dwindled. Guffey refused to become a ghost town, however. Cattle ranching helped sustain the town. A few old buildings, including a small jail, remain today in the peaceful valley community.

WEBSTER

Location: 3 miles west of Grant on U.S. Highway 285

Webster was a rough, tough town. Brawls and shootings were commonplace. When miners and railroad workers were mixed together there was often trouble. North of the townsite are two cemeteries — one for the "citizens" of Webster, and the other a "Boot Hill" for derelicts, drifters, unknowns, and those unfortunates who were "gunned down".

In his Grip-Sack Guide of Colorado, George Crofutt describes the poor predicament a traveler might have while staying at Webster's only hotel. Floorspace large enough for one body, including a blanket, rented for $1.00 per night. When customers exceeded the number of blankets, the blankets were removed from sound sleepers and reused as often as possible. The town was a fairly raw place.

During the late 1870s Webster was the terminus of the Denver, South Park & Pacific Railroad. From this point wagons and stages carried freight and passengers over Kenosha Pass to South Park, or along the North Fork South Platte River and across Webster Pass to the Snake River and Peru Creek mining districts. Railroad fare from Denver was $7.00 one way.

The town is named for William & Emerson Webster who in partnership with the Montezuma Silver Mining Company constructed a road from the mining towns of Montezuma and Saints John across the pass which bears their name, and down into Handcart Gulch and the Hall Valley Mining District.

A long row of coke ovens at one end of town supplied coke for smelters in the Hall Valley and Geneva Mining districts.

Once the railroad topped Kenosha Pass and extended into South Park, the importance of Webster diminished. It was easy for travelers to find more convenient connections and better accommodations elsewhere.

A wagon awaits repair in early Fairplay. *Colorado Historical Society.*

Buckskin Joe in 1864. *Colorado Historical Society.*

Montgomery as photographed by G. D. Wakely in 1864. *Colorado Historical Society.*

Rolling out an ore cart at Alma. *Collection of Dave Southworth.*

The hose crew poses during 4th of July festivities at Alma. *Colorado Historical Society.*

This is the earliest known photograph of Jefferson, taken when it was still a tent colony. The Denver, South Park & Pacific Railroad can be seen in the background. *Denver Public Library, Western History Department.*

The Denver, Leadville & Gunnison Railway Engine No. 192 with its crew and passengers during a stop at Como. *Colorado Historical Society.*

Leavick, looking northwest toward Mount Sheridan. *Colorado Historical Society.*

Balfour in the late 1890s -- a town which came and went in a hurry. *Denver Public Library, Western History Department.*

The railroad depot at Webster in the 1880s. *Colorado Historical Society.*

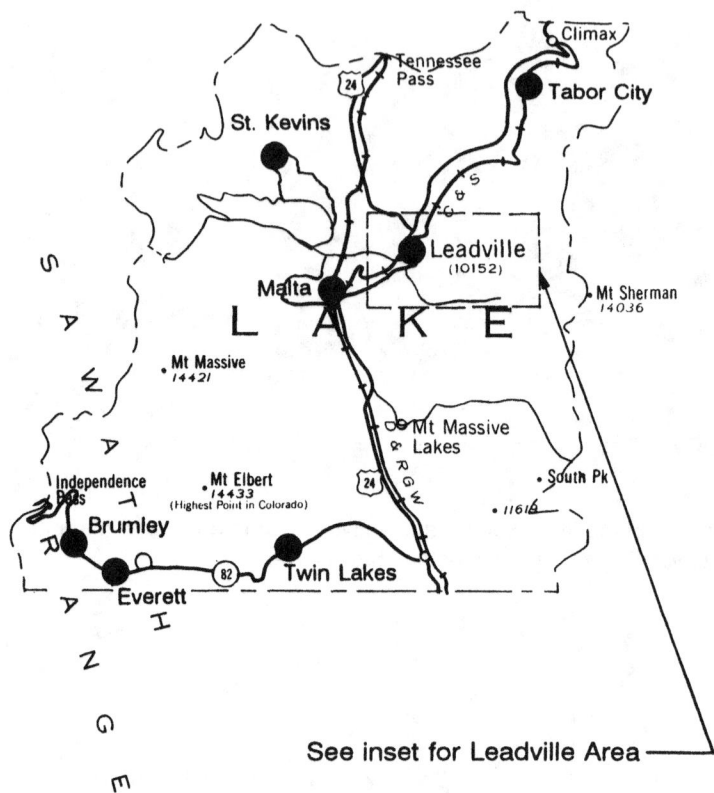

LAKE COUNTY

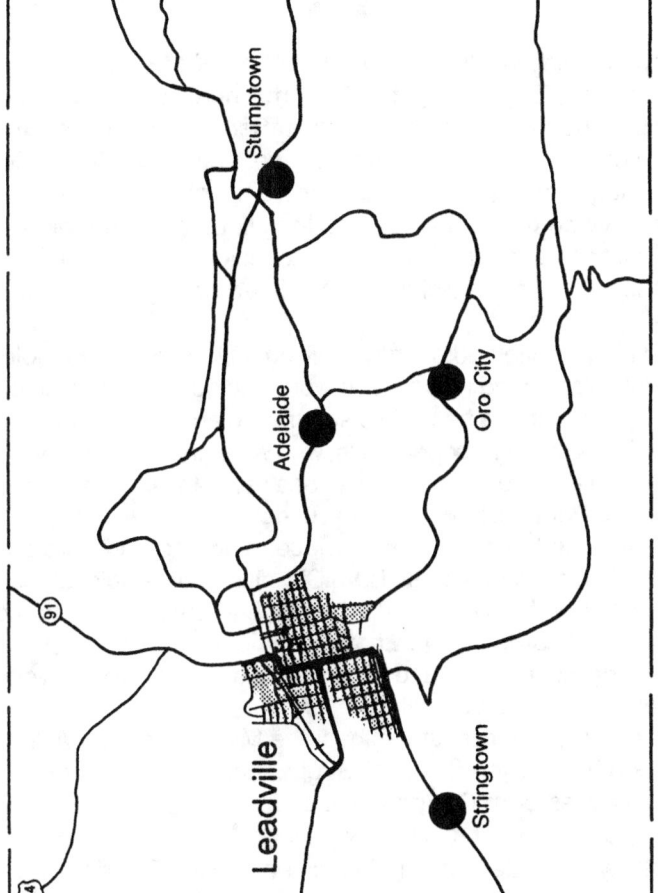

LEADVILLE AREA

"If the food at the camp is a little weak tastin', the coffee's probably strong enough to average it out."

ORO CITY

Location: 3 miles southeast of Leadville via East 4th Street and California Gulch

While prospecting in California Gulch on April 26, 1860, Abe Lee discovered gold and proclaimed, "Boys, I've got all California here in this pan." The stampede was on. Thousands of fortune seekers swarmed into California Gulch and the surrounding mountains. Tents and wagons were scattered from one end of the gulch to the other. Merchants sold supplies from their wagons. Canvas saloons popped up everywhere. Prostitutes had long lines outside their tents. Finally some order began to appear. Tents began to "colonize." The original site of Oro City, the largest camp in the vicinity, was located on the south edge of present day Leadville.

California Gulch yielded over five million dollars in surface gold during the '60s. The heavy black sand was hard to work and good accessible ore began to dwindle. Many prospectors became discouraged and moved on.

Charles J. Mullen and Cooper Smith, who were grubstaked by a Philadelphian named J. Marshall Paul, found a new rich lode at the Printer Boy Mine, and a new boom began. The Printer Boy was the first underground gold mine in the Leadville area and marked the transition from surface mining to underground hard rock mining. The remaining inhabitants of Oro City dismantled their cabins and moved them, along with their furnishings, up the gulch to the new location of Oro City, commonly called "Slabtown." New hotels, saloons, and stores sprang up. One of the latter belonged to H.A.W. and Augusta Tabor. Tabor would eventually make a fortune, then lose it (see Leadville).

Two of the area's successful mines, the Minnie, and the A.Y. (discovered by A.Y. Gorman) helped spring Meyer Guggenheim on his way to a fortune. The mines enabled Swiss-born Guggenheim and his sons to compile much wealth which they spread into other mining endeavors.

Productivity again declined during the early '70s. The discovery of silver in 1875 created another important boom in California Gulch.

Oro City was devastated by the silver crash of '93, but survived past the turn of the century.

Leadville's Carbonate Concert Hall advertised: "Wine, Women and Song — These three are supposed to make life palatial and while nothing of an improper character is permitted we can furnish all three any night."

LEADVILLE

Location: 33 miles north of Buena Vista on U.S. Highway 24

After Abe Lee discovered gold near the site of present-day Leadville (in April 1860), he and his group (there were originally fourteen miners) staked nearly the entire California Gulch with "speculative claims" in an effort to grab all of the property which might be potentially profitable. The plan didn't work, however, for new prospectors demanded a miners' meeting, from which came the official "Bylaws of California Mining District, California Gulch, Arkansas River." They were adopted on May 12, 1860, and established provisions as to the number, size and type of claims which could be filed. Fair laws invited more newcomers. By mid-summer there were 5,000 people scattered throughout California Gulch (also see Oro City).

Among the earliest arrivals into the region that spring, were Horace Austin Warner Tabor and his wife Augusta. H. A. W. Tabor, who was to become a pillar in the development of Leadville, arrived into California Gulch during a food shortage. He sacrificed his oxen so that hungry miners could eat, and by doing so immediately became the friend of many. Augusta established a little store where she sold her baked goods and provided meals as well. Like most of the men, Horace prospected the surrounding hills.

California Gulch yielded over $5,000,000 in gold during the next five years. By 1866, however, the placer gold was gone and most of the miners had packed up and left. Underground mining, which began with the discovery of the Printer Boy in 1868, created a new but modest flurry of activity. Heavy black sand covered the region and made gold mining difficult. In 1875 William Stevens and Alvinus Wood decided to have the sand assayed. They discovered that it contained fifteen ounces of silver per ton and was rich in carbonate but kept their find a secret for nearly two years. When the word spread in late 1877 the silver boom was on.

In 1878 Leadville was a simple community of log structures. By June of the following year, however, it had blossomed considerably. According to Cass Carpenter, as of May 1, 1879, Leadville had "... 19 hotels, 41 lodging houses, 82 drinking saloons, 38 restaurants, 13 wholesale liquor houses, 10 lumber yards, 7 smelting and reduction works, 2 sampling works for testing ores, 12 blacksmith shops, 6 livery stables, 6 jewelry stores, 3 undertakers, and 21 gambling houses where all sorts of games are played as openly as the Sunday School sermon is conducted." Additionally, there were 36 brothels. As George E. King and other architects arrived, Leadville took on an air of elegance and sophistication. This was something the "suburbs" such as Finntown, Stringtown, Stumptown, and

Tintown never had.

In April 1878, Horace Tabor grubstaked August Rische and George Hook to $17 worth of supplies for a third interest in their findings. He added another $47 worth of tools to the grubstake. Rische and Hook discovered the Little Pittsburg (often spelled Pittsburgh) Mine and made Tabor an instant millionaire. Tabor once purchased a "salted" mine, the Chrysolite, but instead of considering it a bad deal, he further developed the property and struck rich silver ore. Later, Tabor purchased the Matchless Mine for $117,000 — possibly the only investment he made without partners. During the peak years of its operation the Matchless yielded $1,000,000 per year.

Augusta Tabor was accustomed to a modest lifestyle, and did not agree with the lavish tastes that accompanied Horace's newfound fortune. He spent or gave away his money almost as fast as he received it. Horace and Augusta drew further apart. It wasn't long before the beautiful divorcee, Elizabeth McCourt Doe, caught his eye. After divorcing Augusta, Horace and Baby Doe were married in 1883 at a huge wedding in Washington D.C., which was attended by many celebrities including President Chester Arthur. Unlike the first Mrs. Tabor, Baby Doe was extravagant and helped Horace spend his money. The silver crash of 1893 depleted the fortune of H. A. W. Tabor. The man who had been mayor of Leadville, Lieutenant Governor of Colorado, and briefly a U.S. Senator, died a pauper in 1899.

Baby Doe's devotion to the "Silver King" was unwavering, even through the tough years. Baby Doe spent her remaining years living in the small shack adjacent to the Matchless Mine, until her frozen body was found one day in 1935. Seldom has anyone experienced such contrasts in beauty and decay or wealth and poverty.

Most discussions of Horace and Baby Doe Tabor and the Matchless Mine include a quotation which was supposedly Horace Tabor's last command (on his deathbed) to Baby Doe — "Hold on to the Matchless." Historian, Caroline Bancroft confessed to the Colorado Historical Society prior to her death, that she had created the quotation stating that she thought it was "good press."

At age 19, Margaret Tobin married J. J. Brown in 1886. Brown, who was superintendent of the Louisville Mine, later struck it rich on the Little Jonny Mine and made a wealthy woman of Maggie, as she was known in Leadville. After the couple moved to Denver, Mrs. Brown attained national fame as a survivor of the Titanic. Her life was further immortalized on stage and screen as, "the Unsinkable Molly Brown".

Many other people of fame or fortune, or both, left their mark on Leadville — and vice versa. Meyer Guggenheim reaped a bonanza from the A. Y. Mine. He invested into the smelting industry and other properties in which his seven sons all became millionaires as well. David May opened the Great Western Auction House and Clothing Store in Leadville on January 1, 1878. This was the forerunner of the department store chain known today as the May Company. The successful hardware store of Charles Boettcher was the foundation for another fortune. Boettcher moved his holdings to Denver and amassed extraordinary wealth. James Viola Dexter was another man made rich by Leadville silver. A log cabin which Dexter fixed up as an exclusive private poker club now sits on property of the Healy House Museum. Dexter was a collector of curios and rare coins. His coins alone were worth over $100,000, a tidy sum in the 1800s. Samuel Newhouse eventually became a copper magnate after he struck it rich in Leadville. Democrat Alva Adams became governor of Colorado three different times (1887, 1897, 1905). The Morning Star Mine added to the riches of two-time governor John L. Routt. The wealthy John F. Campion (also see Twin Lakes) later developed the beet sugar industry in Colorado. Soapy Smith (see Creede) learned the principles of the old shell game in Leadville. Col. William F. Cody, "Buffalo Bill", came to town with his Wild West Show. During his visit he paid tribute to the memory of an old friend, John B. Omohundro, "Texas Jack". Famous gunfighter and gambler, Doc Holliday, made at least two visits to Leadville. Prostitute and storyteller, Laura Evans, spent about three wild years in Leadville. During a tour of mining camps in 1877, Susan B. Anthony stopped in town to lecture for woman's suffrage. While visiting 24 camps Miss Anthony was only able to collect $165, a reflection of the unpopularity of the issue. Several men of the cloth left their mark as well, including Father John Dyer, Father Machebeuf (who made annual visits to California Gulch in the early days), Pastor Arthur Lake, Father Robinson, and the Reverend Tom A. Uzzell. Frank W. DeWalt was president of the First National Bank of Leadville before it closed in 1884. He wasn't much of a president — nor was he much of a gambler. Evidently DeWalt gambled away $50,000 and financed the bordello of Madame Purdy with depositors' money. His wrongdoing netted him a prison term.

Madame Vestal, who operated a dance hall on State Street, was "queen" of the lady gamblers. "Broken Nose" Scotty sold his claim for $30,000 while he was in jail for drunkenness. With the money, he paid the fines for each of the other inmates, bought them new clothing, and wined and dined them before they all wound up back in jail for disturbing the peace. George Fryer discovered silver on the hill which bears his name, then spent a half-million dollars as fast as he was able. When his money ran out, he committed suicide. John Morrissey became rich and famous, but his fame was attributed to his illiteracy. According to one story, he hollered down his mine shaft, "How many of you are down there?" The reply was, "Three." Morrissey yelled back, "Well, half of you come up and have a drink." The *Leadville Chronicle* had a colorful newsman named Orth Stein. Many of the legends and tall tales which emerged from the region were created by Stein. The Maid of Erin Mine made Jack McCombe a wealthy man. He spent much of his wealth, however, sending presents to everyone he knew in Ireland, his home country. One of those who made a name for himself because he didn't strike it rich was James Fenton. After working twelve unsuccessful claims, the dejected miner buried himself with a dynamite blast. Leadville was touched by many others, who were touched by Leadville as well.

In 1878, prospectors, merchants, and others trickled into town from all directions. Dr. David H. Dougan closed his office in Alma, crossed Mosquito Pass, and set up shop in Leadville. Nobody seemed to need a doctor, however, and Dougan sat in his office for 28 days without a customer. On the 29th day he received word that there had been a mine accident and he was needed immediately. That was the beginning of his illustrious career in Leadville. Dougan was a successful physician, then became mayor in 1881, and later was president of the Carbonate National Bank.

A story is told about Tom Lavery, a miner who handled a gun fairly well. Lavery staked a claim and proceeded to work it. The problem was — its location was completely surrounded by property owned by H. A. W. Tabor. Tough Irishman Martin Duggan was marshal of Leadville during the late 1870s and early 1880s. After Tabor was unsuccessful in getting Lavery to leave the claim, Duggan and his deputies were summoned to force Lavery from the property. Lavery wouldn't budge, remaining in the mine shaft with only his head exposed. A gunfight erupted and Lavery shot down two deputies. Lavery in turn was killed in his own shaft.

As previously mentioned, Leadville was once a village of log structures. George Albert Harris built the first log hotel on Chestnut Street in May of 1877. On January 14, 1878, the community was officially named. Beginning with the boom of 1879, the growth of Leadville was rapid. Harrison Avenue emerged as the center of the business community, where most of the hotels, banks and restaurants were located. Horace Tabor, who became mayor in 1878, built both the Clarendon Hotel and the elegant Tabor Opera House the following year. Many important people and famous stars graced the stage of the Tabor Opera House. Lectures were held by British critic and poet, Oscar Wilde. Wilde raised a few eyebrows by drinking several Leadville miners under the table. Central City's Jack Langrishe and his players performed for the more cultured. The great Harry Houdini performed his magic on stage. Shakespearian actor Laurence Barrett appeared with his supporting cast. Boxing matches were held headlined by prizefighters John L. Sullivan and James Corbett. Sousa's Marine Band was among many to appear at the facility. The elaborate Tabor Grand Hotel, with silver dollars set into its lobby floor, was opened in 1885.

Most of the many saloons, gambling dens, dance halls, and brothels were located along State Street and in its dingy corridors. There was Tiger Alley, French Row, Coon Row, and Stillborn Alley. It was one of the most wicked and rowdy areas in the entire old west. There were elegant bordellos and one-girl cribs. Many of the dance halls and saloons had rooms above where more than just a garter was removed. Tough men, painted women, and slick gamblers could be found at "joints" such as the Bucket of Blood, the Pioneer, the aptly named Red Light Hall, the Carbonate Concert Hall, the National, the Odeon, the Bon Ton, the Bella Union, and the Little Casino. Madams such as Mollie Price, Mollie May and Sallie Purple ran "houses" which catered to a fairly respectable clientele. In contrast, the cribs had girls of every size and shape, and every color and origin, and would cater to anyone. Top prostitutes such as the Pioneer's sassy Maude Deuel could make as much as $200 per week. The Texas House, which was located on the corner of State Street and Harrison Avenue, was one of the largest gambling halls in Colorado. Many fortunes were made in Leadville —

and a part of many were spent on State Street.

For the most part, justice in early Leadville left much to be desired. Murder and thievery were common place. Vigilantes often took the law into their own hands. On one occasion three Chinese workers were hanged, then shot to pieces and dumped into an empty prospect hole. Sometimes, the "guilty" were hanged and left swinging for days as an "example" to others. The first person ever jailed for insanity was so judged because he spent all his time in prayer. It was dangerous for an individual to walk the streets at night, even if he had a cocked pistol in hand. Armed guards were stationed at the Presbyterian Church construction site to discourage claim jumpers. Becoming a marshal in Leadville sometimes meant signing your own death warrant. And many a man was shot for a gambling debt, a dispute over a claim, in an argument over a woman (and there weren't many for a while), during a drunken brawl, or for no reason at all.

Like Tabor, most of the Carbonate Kings lost their fortunes in 1893. President Cleveland repealed the Sherman Silver Purchase Act whereby the government would no longer purchase silver to back its currency. The foreign market for silver collapsed as well. Leadville was dealt a crippling blow, but it wasn't about to die.

The economy of Leadville has rebounded magnificently. In 1895, silver production exceeded every year in Leadville's short history with the exception of the boom year of 1880. the total production of area mines was also the greatest since 1889. Once more, money was flowing.. Seeking to bring tourists and new money into the community, several businessmen decided to create a new and different attraction — an ice palace. The structure looked much like a medieval castle but was constructed with 5,000 tons of ice. As many as 250 men were hired to cut the ice and build the palace which covered much of the five-acre site. Within the confines of the five-foot thick walls, were a skating rink which was 80 feet wide by 190 feet long, a grand ballroom, and a large dining room. The palace officially opened on January 3, 1896, amid much fanfare. City politicians, civic clubs, two thousand members of the Miners' Union, the Dodge City Cowboy Band, and others participated in a huge parade to commemorate the occasion. It was necessary for each of the three railroads serving Leadville to add extra passenger cars to scheduled trains in order to accommodate the throngs of tourists flocking to see the unique structure. Because of an unusually mild winter, the ice began to melt prematurely. The attraction permanently closed on March 28, 1896.

Silver mines maintained an annual average production of about $10,000,000 for many years. Lake County ranks first in the state in the production of silver, lead, and zinc, second in copper, and fourth in gold. No visit to the area is complete without a tour of the old mining remnants outside Leadville. Within the community — which is proud of its heritage — there are many historical attractions to remind one of the glorious past.

STRINGTOWN

Location: 1 mile southwest of Leadville on U.S. Highway 24

South of Leadville along what is now U.S. Highway 24, stretched a group of "suburbs" once known as "Smelter Valley." Jacktown (which had a bowling alley), Stringtown, Bucktown, and Little Chicago (across the river) were blue-collar towns. In sharp contrast to some of the Victorian gingerbread in Leadville, Stringtown and its neighbors were comprised of many tenement shacks and plain cabins mixed with a few commercial buildings. The nearby chimneys at the smelting furnaces noted their main source of prosperity.

The giant Arkansas Valley Plant, a lead smelter, sprawled out between Stringtown and Bucktown. Most of Stringtown's residents were Arkansas Valley workers, as were those of Bucktown and Little Chicago. Founded in 1879 as the Billing and Eilers Smelter, the Arkansas Valley Plant was operated by the American Smelting & Refining Company. Several other smelters operated in the vicinity.

Stringtown had a hotel — the Great Northern. There were many saloons, and cribs occupied by girls of the night — and day. The valley was a tough place, with fights and violence commonplace.

Today "Smelter Valley" is marked by the massive slag piles from the Arkansas Valley Plant and other smelters, a few ruins, and industrial sites still in use.

ADELAIDE

Location: 2 miles east of Leadville via East 5th Street and Stray Horse Gulch

In 1876 the Adelaide Mine was located in Stray Horse Gulch. The town of Adelaide grew up in the flat valley near the mine of the same name.

Although Adelaide was only two miles from Leadville and the route looked like one continuous town, the "suburb" had its own identity. The town had a post office, school, large smelter, and twenty-eight commercial buildings which included several saloons and stores. By 1879 Adelaide had a population of about 1,000.

There were some good mines in the area which stayed busy through the silver boom. In addition to the Adelaide Mine, the Eureka, Humboldt, Dolemite, and Morning Glory all produced well. A short distance above Adelaide was the location of the famous Little Jonny (sometimes misspelled as Little Johnny) and the other Ibex mines. The Ibex Camp served the several properties of the Ibex Mining Co..

Two of the mines in the vicinity had a problem keeping workers. Senator Gallagher's Mikado and Camp Bird (not to be confused with the one at Ouray) were said to be haunted. A ghost appeared, so they say, and miners quit. Gallagher was killed in the Moyer Mine near Oro City. According to legend, his spirit made the rounds of his mining properties on a regular basis.

When the silver market became rocky, gold producers like the Black Prince Mine helped sustain things. For the most part, Adelaide followed the same pattern of deterioration as did the rest of Leadville's suburbs.

STUMPTOWN

Location: 3 miles of east of Leadville via East 7th Street and Little Stray Horse Gulch

Stumptown was originally named Stumpftown for Joseph Stumpf, a founding father. The community was a "suburb" of Leadville and very similar to most of the mining camps in the vicinity. Mines sprinkled the area east of Leadville like pepper from a pepper shaker. Wherever there was a cluster of mines, a camp usually sprang up, and some were very close together. Such was the case with Stumptown.

The campsite is located in South Evans Gulch a half-mile east of Evansville. Like its neighbor down the gulch, Stumptown had several saloons and other businesses — the most renowned being a crowded pool hall, where the betting was usually heavy.

There were many productive mines in the vicinity. The Boulder, St. Louis, Louise, Winnie, Ollie Reed, Favorite, Little Bob, and the Miner Boy were some of the best. In the 1890s, a squabble broke out around the St. Louis Tunnel. The Miner Boy and the Colorado Prince had a dispute which wound up in violence. Both mines were burned. Cooler heads finally prevailed and the claims were consolidated and worked by the St. Louis Tunnel.

Stumptown began in 1879 when carbonate ores were discovered in the Little Ellen Mine. The camp boomed until the silver crash of 1893. It experienced a rebound in 1895 but began a gradual decline thereafter. The site, located in a shallow valley, was abandoned by the late 1930s.

TWIN LAKES

Location: 21 miles southwest of Leadville via U.S. Highway 24 and State Highway 82

In the shadow of Mt. Elbert, Colorado's highest peak, the mining camp of Dayton was established in 1863. Just a short distance from Dayton, the Ryan House was constructed during the same year. It is the oldest stagecoach stop in Colorado. Dayton, which was the county seat for almost two years, eventually phased into the area around the Ryan House to become the resort town of Twin Lakes.

Concord stages made the regular run from Leadville to Twin Lakes. Passengers traveling to the boom town of Aspen would transfer to a canvas-top stage for the grueling run over Independence Pass, known then as Hunter's Pass. The stage road ran through Twin Lakes one block north of present Hwy 82. The Ryan House, constructed of logs, has been renovated and now is an attractive house adorned with white shutters.

Of several hotels, the Inter-Laken was the most popular. It featured a luxurious dining room, ballroom, and gaming hall. The Inter-Laken was expensive — rooms renting for as much as four dollars per day.

Leadville mine owner John Campion constructed a magnificent palace facing the lakes. It was lavishly furnished with European finery. Many were elegantly entertained at Campion's Lodge.

Tourists always outnumbered residents, of which there were about 200 during the peak years. Many old buildings remain in Twin Lakes — which is inhabited throughout the year.

MALTA

Location: 4 miles southwest of Leadville on U.S. Highway 24

Rich ore containing silver and lead was accidently discovered in 1871 by three men trying to find a mountain trail. The men were disappointed because only a trace of gold was found in the ore. Nevertheless they went to work on their claims, and the Homestake region was born. A small smelter was built in 1875 to process the ores from the region. The site, named Swilltown, was later to become Malta. The Swilltown smelter was the first to handle silver in the whole Leadville area.

In 1879, at a cost of $5,000, a racetrack was constructed at Malta. Its horse races were popular and drew a fine crowd from Leadville and the surrounding area.

Alcoholic beverages were important to the economy — there were many saloons and also a brewery. The community peaked about 1881 with a population over 300. During the ensuing years, larger and more efficient smelters were constructed closer to Leadville. The railroad helped sustain some business, but Malta became less significant. Today, the charming red school house, topped with it's bell tower, marks the location alongside U.S. Highway 24.

According to Orth Stein of the <u>Leadville Chronicle</u>, the frozen body of a female prospector was found north of Saint Kevins near Homestake Peak. The body was carried to the newspaper office and placed on a desk near a pot bellied stove. Newsmen watched as the women thawed out — limb by limb. Once completely thawed, she sat up, then walked out of the office — never to be seen or heard from again.

SAINT KEVINS
Location: 7 miles northwest of Leadville

Realizing he would never become rich working as a carpenter for the Colorado Central Railroad, Thomas Walsh turned to prospecting. His diggings carried him to many areas, and eventually to the spot northwest of Leadville where he made a good silver strike. In the mid '80s he, and others, platted the town of Saint Kevins north of the area which is now Turquoise Lake. There were several productive mines in the area. The Saint Kevins, Griffin, Amity, and President all mined the silver, lead, and copper deposits in the area. There was even a little gold. A Stamp Mill was constructed at the camp.

Saint Kevins had a boardinghouse, school, and mercantile store — but the camp never grew very large. Most of the inhabitants left following the silver crash of 1893. A little mining continued after the turn of the century.

As for Thomas Walsh, the crash of '93 broke him as it did many. What he lost, however, was a drop in the bucket compared to the fortune he would soon realize from the famous Camp Bird Mine near Ouray.

TABOR CITY
Location: 10 miles north of Leadville on State Highway 91

Tabor City (sometimes simply called Tabor) was a short-lived mining camp and supply station north of Leadville on the road to Fremont Pass. The site was purchased by a Colonel Taylor in 1879, who platted a town and called it Taylor City. Some confusion exists with regard to this name. Across Buckeye Peak to the west, near Tennessee Pass, there existed another camp named Taylor City which also had a very brief life. Perhaps because of the confusion, the camp residents changed the name to Tabor City in honor of Leadville mayor Horace Tabor — who had just become a millionaire by selling his interest in the Little Pittsburg Mine.

By 1880 Tabor City had a population of about 150. There was a general merchandise store, two hotels, two restaurants, a blacksmith shop, livery stable, and many cabins.

Miners had high expectations but low realizations. Rich veins were never found — only low-grade ore. By 1881 nearly everybody had left and the post office was discontinued.

They say the most exciting thing to ever happen around Brumley was when two stagecoach drivers fought over the same girl. One stabbed the other. We still don't know who wound up with the girl — the stabbed or the stabber.

BRUMLEY
Location: 10 miles west of Twin Lakes on State Highway 82

Amidst productive mining on Mount Champion, Star Mountain, and in Mountain Boy Gulch, blossomed the camp of Bromley Station — named for the newly constructed hotel and its owner. The name was shortened to Bromley, and then mysteriously became Brumley.

Of the mines in the area, the best producer was the Mount Champion, so dubbed for the mountain on which it was located. A three mile long aerial tram was constructed at a cost of $115,000.00 to lift ore over the top of the peak and down to Halfmoon Gulch where it was hauled off by wagon.

The town was never very large but was an important stage stop. The Independence Pass Road was a toll road, and Brumley served as a tollgate. As travel over the pass declined, so did the settlement. With the exception of two parallel concrete footings and ruins of some cabins, Brumley is gone.

EVERETT
Location: 7 miles west of Twin Lakes on State Highway 82

The Halfway House was an important stagecoach stop on the road over Independence Pass — the road from Leadville to Aspen. C. M. Everett platted a town and named it after himself. In addition to several stores and many cabins there was a two-story stage station and hotel — the Everett House.

C.M. Everett built two stamp mills at the confluence of North Fork and South Fork of Lake Creek to crush ores packed in from the Ruby Mine and others.

When the railroad arrived in Aspen, the pass became much less useful. As wagon and stage travel declined, so did mining — and so did the town of Everett.

An old cabin ruin and foundations mark the spot where Everett once stood.

Because of its high elevation, Leadville was dubbed the "Cloud City".
Colorado Historical Society.

Wolfe Londoner, Leadville's first real merchant, arrived in 1860.
Colorado Historical Society.

Leadville's Tabor Opera House opened to the public on November 20, 1879. The bridge crossed over St. Louis Avenue to the Clarendon Hotel rooftop. *Colorado Historical Society.*

The Inter-Laken Hotel at Twin Lakes was operated by James V. Dexter after 1895. *Colorado Historical Society.*

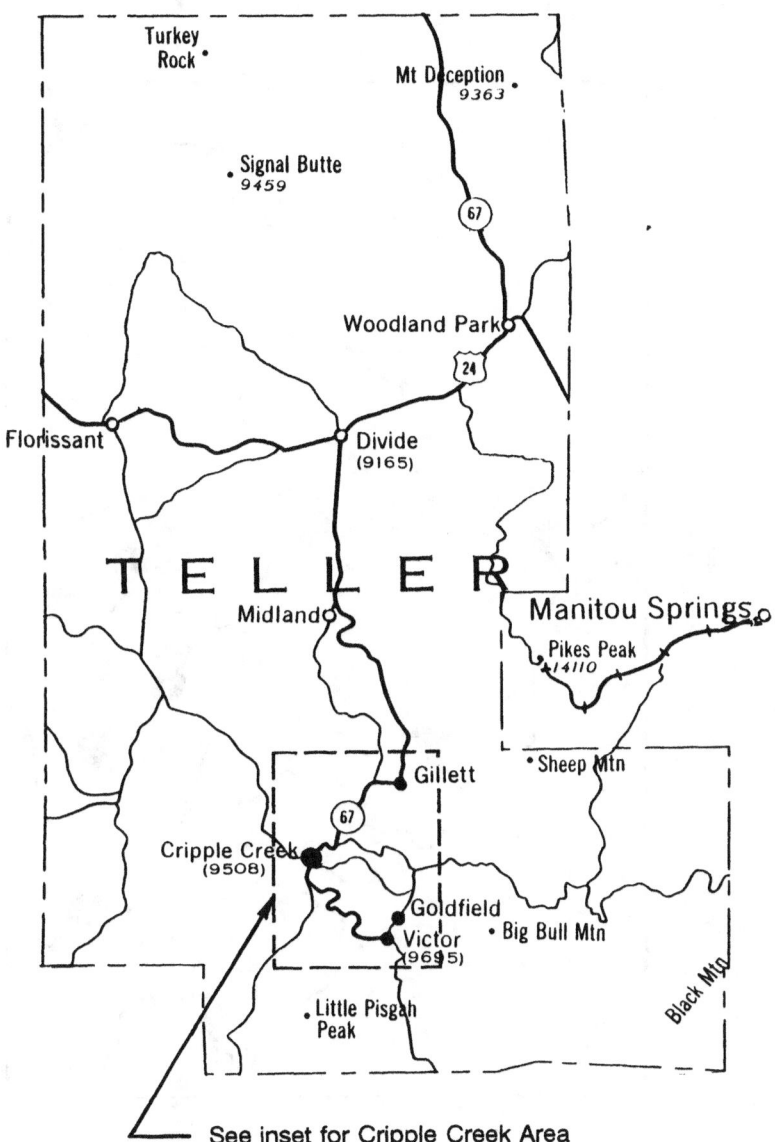

TELLER COUNTY

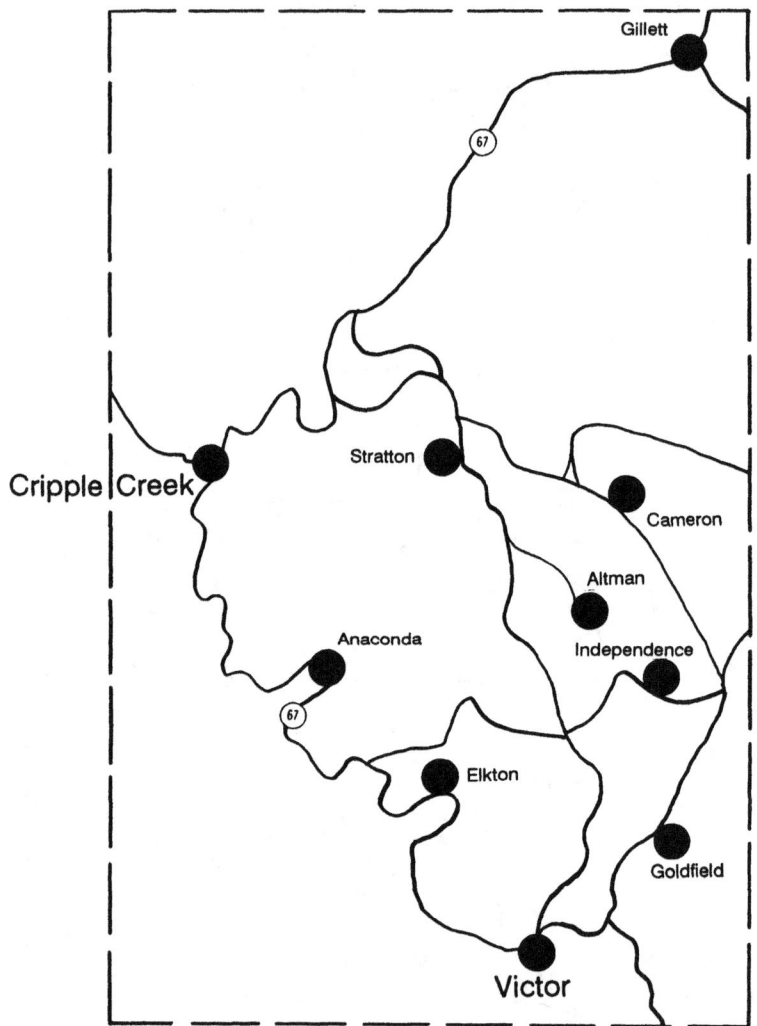

CRIPPLE CREEK AREA

The "purple mountain majesty" of Pike's Peak inspired Katherine Lee Bates to write "America the Beautiful."

CRIPPLE CREEK

Location: 43 miles west of Colorado Springs via U.S. 24 & State Highway 67

In 1890 Bob Womack discovered gold in Poverty Gulch. He sold the claim for a mere $500. It later became part of the Gold King mining properties which yielded $8,000,000. Two Denver cattlemen, Horace Bennett and Julius Myers, purchased the land only to find a couple of colonies blossoming on their property. A cluster of tents at Hayden Placer cropped up near a mixture of tents and small log cubicles in a community already named Fremont. Quick to realize what was happening, Bennett and Myers promptly platted a town and sold lots. The two main streets were named for themselves, and the town of Cripple Creek was established.

Bennett Avenue was lined with hotels, department stores, specialty shops, doctors' offices, and many other kinds of businesses. The stock exchange and half of the banks in the district were on Bennett Avenue. It was the center of high-finance. Most of the "low" finance took place on Myers Avenue. Parlor houses, cribs, gambling dens, dance halls, and many saloons lined the street which became one of the most notorious red-light districts in the West.

On April 25, 1896, a fire broke out in a Myers Avenue parlor house. Aided by a strong wind, it swept through the lower end of town, destroying about one third of the business district in just four hours. Just four days later, on April 29th, a second fire which began at the Portland Hotel engulfed most of what remained. The two fires left thousands homeless and destroyed over $2,000,000 worth of property.

Help came from far and wide — especially Colorado Springs. Tents and supplies poured in to house and feed the homeless. Winfield Scott Stratton picked up the tab on many of the relief items. There wasn't time to mourn the losses. Construction began immediately to build a "new" Cripple Creek of brick and stone — thanks to the banks of Colorado Springs.

The National Hotel, an elegant five story structure, housing 150 rooms, was constructed on Bennett Avenue. The hotel had elevators and its own electrical system. Next to it was built the Gold Mining Stock Exchange. At the corner of Bennett and 3rd stood the new May Company department store. Attractive new buildings lined both sides of Bennett Avenue.

On 3rd Street, just north of Bennett Avenue, the Imperial Hotel was built (the only one of Cripple Creek's original hotels still standing). Myers Avenue also rebuilt. There was a carnival-like atmosphere. The Bon Ton and other variety theaters had three or four piece bands playing out front in order to attract customers to their establishments. The Old Homestead at 353 Myers Avenue was a posh brothel with crystal chandeliers and Oriental carpets. The establishment (which still stands) had a select clientele. Other fancy houses on Myers Avenue were the Royal Inn, the Boston, Nell McClusky's, and Laura Bell's. Myers Avenue tailed out into Poverty Gulch where most of the one-girl cribs were located. The Last Chance, the Miner's Exchange, the Dawson Club, and the Opera Club were a few of the seventy three saloons in Cripple Creek at the turn of the century. The largest gambling hall was Johnny Nolan's at 3rd Street and Bennett Avenue.

The night life in Cripple Creek was — to put it mildly — wild! A tombstone in Mt. Pisgah cemetery bears the inscription: "He called Bill Smith a liar." One night, after a man was shot at the Dawson Club, a witness stated, "as the man lay dying, some of the crowd urged him to the bar for a drink." The town had a busy undertaker.

Culture in Cripple Creek was progressive. Famous performers of the day graced the stages of both the Grand Opera House and the Butte Opera House. The Crystal was a popular vaudeville theater. The opera houses and theaters had many uses, including town meetings and sporting events. The Grand Opera House would fill to capacity to watch such prizefighters as former heavyweight champions Jack Dempsey (who once worked at the Portland Mine) and Jack Johnson. There were many lodges, fraternal organizations, women's clubs, and other civic groups in Cripple Creek. The schools were considered excellent. The high school had night classes for adults.

Dubbed "the world's greatest gold camp" the Cripple Creek Mining District was the fifth largest single gold producing area in the world. The total gold production for the district has been estimated at a half-billion dollars. Many millionaires were created. Although many of the largest and richest mines were located near Victor, most of the mine owners lived in Cripple Creek. Cripple Creek reaped the profits. Closer to home, the C.O.D. Mine in Poverty Gulch made a wealthy man of Spencer Penrose. Penrose's brother once wired him $150 so he could purchase a railroad ticket back to Philadelphia. Penrose invested the money and later paid his brother $10,000 which was the return on his $150 investment. Penrose, who constructed the Broadmoor Hotel in Colorado Springs, profited from several mining and milling operations.

Two electric trolley systems linked Cripple Creek and Victor. The Low Line trolley ran every thirty minutes via the communities of Anaconda and Elkton. The High Line used a different route through the little town of Midway. The Cripple Creek Mining District was the only place workers could ride to the mines and return by way of trolley.

The narrow guage — the Florence and Cripple Creek Railroad — arrived in 1894. The line, which ran until 1912, paid for itself during its first year of operation. The Midland Terminal Railway, a standard guage, arrived in 1895. It ran over a half century until 1949. The Colorado Springs and Cripple Creek District Railway, known as the Short Line, began operating in 1901 from Colorado Springs through a wonderfully scenic area to Cripple Creek. The spectacular scenery along the way was once described by Teddy Roosevelt as "the trip that bankrupts the English language." The Short Line discontinued service in 1920.

Sometimes there is a great disparity between population estimates and U. S. Census figures. Every town in the district peaked about 1900, a census year. Officially the population of the town of Cripple Creek was 10,147. One source claimed it was 50,000 — while statistics showed the whole district about 30,000. Census figures in western boom towns were notoriously low, but criteria for some pretty good assumptions.

When one takes into consideration that many of the mines were working three shifts in a twenty four hour period; that hotels were jammed to capacity; that some boarding houses rented cots by "shifts"; and that there was more than one family in some houses — it is easy to understand the inaccuracies. While there were always a number of transients, realistically the population of the town of Cripple Creek was probably under 12,000 in 1900.

Cripple Creek was named county seat of the newly organized Teller County in 1899. There were eight newspapers published in town. Seventy-two lawyers settled claim disputes. Forty stockbrokers guided many investors. Much buying and selling necessitated thirty-nine real estate agents. Cripple Creek was definitely the commercial hub of the district.

The ore dumps and headframes of once productive mines, rimmed by the view of breathtaking mountains, encircles the gold-town nestled in the valley below. Interest in Cripple Creek has grown over the years. The advent of limited stakes gambling in October 1991 has created a new boom which offers a glimpse of the glory days of a century ago. The community has undergone a major face-lift. Traffic has increased 400 percent. Tourists are having an impact on the gaming industry — and it's a sure bet that gaming is having a decided impact on tourists.

"Immigrants heading west took with them hope, memories, and what they could stuff into their wagons."

VICTOR
Location: 6 miles southeast of Cripple Creek

Victor's history was one of fortune and misfortune — in a very plural sense. Gold and glitter were mixed with tension and tragedy.

Winfield Scott Stratton's Independence Mine was the most celebrated in the district. The Portland had the greatest yield, but the Ajax also produced well. On the edge of town was the Strong Mine, another top moneymaker. Battle Mountain, where these four mines were located, produced over $125,000,000 worth of gold, making it one of the richest spots on earth. In the center of town, Frank and Harry Woods discovered the Gold Coin Mine while excavating for a hotel foundation. Gold was found under the athletic field at Victor High School where famed radio newsman Lowell Thomas later graduated. In fact, some people dug gold out of their back yards. It seemed as though gold was everywhere.

The area, originally a helter-skelter tent colony, became organized in 1893 when the Woods brothers purchased a 160 acre tract, once the Mount Rosa Placer property. The Woods Investment Company platted the acreage and sold lots.

Victor became the second largest city in the district — only Cripple Creek was larger. False fronted buildings stood shoulder-to-shoulder throughout the business district. There was much activity and growth was rapid. People clamored for a place to sleep. The boarding houses were packed. Miners stood in long lines at meal times.

At the corner of Victor Avenue and Fourth Street stood the Hotel Victor. Its hotel lobby was a favorite place to meet and was always crowded. A fine city hall was constructed. There were many banks and churches. Saloons were everywhere and many were open twenty-four hours a day.

The Florence and Cripple Creek Railroad arrived in Victor in 1894, followed by the Midland Terminal Railway the next year. The Colorado Springs and Cripple Creek District Railway (commonly called the Short Line) arrived a few years later. There were two trolley lines — the High Line and the Low Line. Trains and trolley cars passed through Victor every few minutes. The city had become the railroad center of the district.

On August 21, 1899, a fire broke out in Paradise Alley (in the red-light district). Nearly four hours later, twelve blocks had been totally destroyed, including many of the buildings of the Gold Coin Mine. Estimates of the damage were huge. Scores were left homeless. The rebuilding began immediately — this time with stone and brick.

A new Victor arose from the ashes. The business district was modern and progressive. The population had reached 7,000, and the city was fifth largest in the state of Colorado. Just a few months before the fire, the Woods brothers had built the Gold Coin Club for their employees. After the fire leveled the original building, a magnificent new Gold Coin Club was constructed. It housed a 700 volume library, bowling alley, ballroom, dining rooms, and game rooms, and also had its own band. The club was the pride of Victor. The red-light district reconstructed its brothels and cribs. The saloons were also rebuilt — there were 37 at the turn of the century. Victor was a wild community.

The city loved to celebrate. Thousands attended the annual Fourth of July festivities. Because it was a strong union town, Labor Day parades were long and joyous. Every time a circus came to town (and there were many) thousands again turned out for the colorful circus parade.

Victor, including its nearby mines and neighboring communities, was the scene of two violent labor wars. In 1894, union officials failed in their efforts to establish a uniform wage throughout the vicinity. The union wanted $3.00 per day for eight hours of work. The mine owners ignored the union's request. A strike was called by the Western Federation of Miners, picket lines were established, and trouble began. Several skirmishes occurred before Governor Davis H. Waite was called in as an arbitrator. He settled the strike, and established a standard wage of $3.00 for an eight hour day, which is what the union wanted in the first place. The Western Federation of Miners called a second strike in 1903 to protest the firing of two men for their union activities in nearby Colorado Springs. Owners reopened the mines with heavily guarded non-union scabs. During the ensuing violence, two mine officials were killed near the town of Independence in an explosion at the Vindicator Mine, a train carrying non-union persons was wrecked, and fifteen scabs fell to their death because someone had tampered with an elevator car at the Independence Mine. Colorado had a new governor who sympathized with the mine owners. Governor James Peabody sent in the militia and established a state of martial law. After six months of occupation and relative quiet, troops were withdrawn. Once again violence erupted. On June 6, 1904, a bomb exploded in the depot at Independence, killing thirteen scabs and injuring many others. The militia was called in again. This time union people were shipped out of the area by rail to Kansas and New Mexico with orders not to return. The labor war was over, but Victor and the mining community never recovered.

Although some mining activity continued until recently, and probably will again in the future, the glory days of Victor were over.

The 1990 census lists the population of Victor at 258. Such is the population that exists amidst many boarded up buildings and closed mines — rememberances of the hustle and bustle of the boom days.

"When a prospector's too old to set a bad example, he hands out good advice."

GOLDFIELD
Location: 1 mile northeast of Victor

James Doyle and James Burns were prospecting on Battle Mountain. Their capital was low, so they took on a third partner, John Harnan. The trio struck paydirt. In 1894, the partners formed the Portland Gold Mining Company. Their mines became the largest producers in the area, yielding $30,000,000 over the next seventeen years.

The town of Goldfield was established at the base of Battle Mountain in 1895 by the Portland owners and was incorporated in 1899. It became the third largest city in the Cripple Creek area with a population of 3,500 at the turn of the century.

Three railroads, the Midland Terminal, the Golden Circle Railroad (part of the Florence and Cripple Creek Railroad), and the Colorado Springs and Cripple Creek District Railway all operated through town. Most of the ore which was mined in the district was shipped from Goldfield.

The community had two newspapers, *The Goldfield Leader*, a daily, and *The Goldfield Times*, a weekly. There were five hotels, two schools, four churches and many stores and saloons. Also, there was one tamale peddler.

Prior to the labor war of 1903, Goldfield was a strong union town. Many union officials were residents and also town officials. When the union was crushed and its leaders driven out of the area, the mine owners governed the city once again.

A few buildings remain from Goldfield's once bustling past, including the city hall. Today, the community has a few permanent residents. As one looks across the grid of streets and old buildings beyond to the mine-scarred hills, it is easy to visualize the activity that once existed.

GILLETT

Location: 7 miles northeast of Cripple Creek

Although it was a mining town and a railroad town, Gillett is most famous for its "recreation." It was known as a bachelor town since so few families were among its residents. The two largest employers in the area were the Gillett Reduction Works and the shops of the Midland Terminal Railway which, combined, employed over a hundred men. The town, with a population of about one thousand at its peak, had many shops and boarding houses and the Monte Carlo Casino, which was eventually converted into the Gillett School. A half-mile racetrack, Sportsman's Park, on the north edge of town was the site of many horse races for several years. The town is best remembered, however, for the Gillett Bull Ring which was constructed in 1895 inside the oval at the racetrack.

Upon completion of the bull ring, J. H. Wolfe and his co-promoters staged a three day "fiesta" August 24 - 26, 1895. The event was highly advertised. Matadors and bulls were imported from Mexico. Tickets sold for $5.00 each day. Whether or not the bulls were tired from their trip to Colorado is not known, but they refused to fight. One was tortured for twenty minutes before it was killed, and the show turned into a sickening fiasco. The promoters expenses of $7,000 exceeded the total receipts of $2,600. The show was a flop.

INDEPENDENCE

Location: 2 miles north of Victor

Independence was named for Winfield Scott Stratton's discovery, the lucrative Independence Mine. A post office was established in 1899. By 1900 the population exceeded 1500 people.

The community is famous for its role in the Cripple Creek Mining District labor strike. In November of 1903 a union terrorist, Harry Orchard, who was allied with the Western Federation of Miners, set off an explosion in the Vindicator Mine killing a mine superintendent and a shift boss. He also tried unsuccessfully to assassinate Colorado's Governor Peabody, the state's Supreme Court Chief Justice William Gabbert, and Justice Luther Goddard. In June 1904 Orchard blew up the Independence railroad depot killing thirteen men and injuring several others. This gruesome incident accelerated the end of the labor conflict. Harry Orchard was arrested and convicted of the murder of a former governor of Idaho. He spent the rest of his life in the Idaho State Prison.

The economy of Independence thrived with the success of the Vindicator and the Hull City mines. The latter property was the primary catalyst which helped the settlement spring to life in 1895. It shipped about 20,000 tons of ore per year during its heyday. Independence was inhabited until the Vindicator Mine shut down in 1959. The remains of many cabins cover the hillside.

ELKTON
Location: 2 miles north of Victor

The Elkton Mine was discovered by a greenhorn prospector who named it for the palmated elk antlers he had found at the site. Half interest in the mine was given to two grocers as payment for a $36.50 bill. The Elkton subsequently produced $16,200,000 (720,000 ounces).

One of the district's largest producers, the Cresson Mine also had a strange beginning. The claim had set dormant for several years. A geologist was asked to investigate the property to see if it had any value. It did, and the Cresson produced $51,200,000 (2,274,000 ounces) before it closed in 1959.

During the year 1915, miners struck a chamber of nearly pure gold — 30 feet in diameter and 20 feet deep — the famous Cresson Vug. The discovery was made at a depth of 1,200 feet, and was so rich that a vault door had to be constructed over the entrance to the chamber. Shipments from the site were accompanied by armed guards.

About midway between Cripple Creek and Victor, the town of Elkton grew around the mine with the same name. It absorbed the nearby camps of Beacon Hill, Arequa, and Eclipse. All three of the railroad lines which serviced the district ran trains through Elkton. The town had plenty of saloons, five hotels, three groceries, and several other stores including a rarity in the virgin West — a bookstore. Elkton's population peaked about 1905 at over 2,500.

The ghost of Elkton is located on private property and is patrolled by mining employees. Permission should be obtained before entering the site.

Today, the Cripple Creek-Victor Gold Mining Company is removing two of the remaining buildings to afford access to new open-pit mining which will be undertaken in the future.

CAMERON
Location: 6 miles north of Victor

Cameron was a small town whose peak population barely exceeded 400. The community (once called Gassy and later Grassy) had a modern school, a fine city hall, a newspaper named *The Golden Crescent*, a few stores, and a wonderful recreational area — Pinnacle Park.

The thirty acre park, which was Cameron's claim to fame, had a stadium which seated 1,000 people. The park, which was constructed for $32,000 by the Woods Investment Company, also had a fine zoo, a dance pavilion, restaurants, and other amenities.

Pinnacle Park was architecturally very attractive. The dance pavilion and each of the buildings had hip-roofs. On square structures, such as towers, the roofs were hip-pyramids. Buildings had exteriors of diagonally cut (at 45 degrees) wood. This look was carried throughout the park, on the pavilion's gables, on a bridge, even on some fences.

It is said that 9,000 persons attended the Labor Day celebration in 1900. Admission was ten cents per person.

There was much mining activity throughout the district, and the area around Cameron was no exception. The Elsmere, Lansing, Wild Horse, and others were good producers for a while.

A decline in the popularity of Pinnacle Park, coupled with less mining activity and much strife created by the labor wars, aided the demise of Cameron.

Violence in Altman was so commonplace that an undertaker once offered group rates for funerals.

ALTMAN
Location: 3 miles north of Victor

Altman, known for a while as the highest incorporated town in the world (even though it wasn't), was one of the larger towns in the Cripple Creek district. The town was named for Sam Altman who operated the first stamp mill and saw mill in the district. The post office was established in 1894.

By the turn of the century, the population of Altman had peaked at about 1,200. The Monte Carlo and the Thirst Parlor headed a list of several saloons.

Additionally, there were two restaurants, a drug store, an assay office, two mercantile stores, and four boardinghouses. The community was very civic minded, having a host of fraternal organizations that met regularly.

The small residential community of Midway became a "suburb" of rapidly growing Altman as it expanded over the hilltop.

The Pharmacist Mine, located on the hill above Altman, was staked out by two druggists who had never done any mining. Supposedly they threw their hats into the air and filed a claim at the spot where they landed. The mine was one of the area's top producers yielding a half-million dollars in gold.

Altman was a stronghold of the Western Federation of Miners and its union leader "King" Calderwood. The notorious Battle of Bull Hill, a labor war, was fought between union members and miners around the campsite in 1894 — prior to the incorporation of the town.

There was much violence in the Cripple Creek district for several years, and each town needed tough lawmen. Altman had a couple of good ones. Legend has it that Sheriff Mike McKinnon gunned down six Texas bad guys in a bloody shoot-out, in which he also was killed. "General" Jack Smith and his notorious Smith Gang had been terrorizing the town of Altman. He was jailed by Marshal Jack Kelly but released on bail. Kelly got word that Smith was waiting for him at one of the local bars. As Kelly walked through the door, Smith shot — and missed. Kelly didn't — killing Smith instantly.

A devastating fire in 1903 destroyed much of the town. Over half of the residents were left homeless.

STRATTON

Location: 3 miles east of Cripple Creek; 5 miles north of Victor

The town of Stratton (also known as Winfield) was headquarters for Winfield Scott Stratton, the Cripple Creek district's first millionaire. It was a company town comprised of a mixture of red brick structures and wood frame buildings. Most of the residents were miners who lived in company boardinghouses.

Stratton, who sold his Independence Mine for $10,000,000 in 1899, had a theory that all of the gold veins in the Cripple Creek area extended from a common "bowl of gold." He built his town as a base of operations to prove his theory. Stratton died in 1902 with his theory unproven. Upon his death, he willed $4,000,000 to the Myron Stratton Home for the handicapped near Colorado Springs. A mining operation tore down most of the buildings in the 1980s.

ANACONDA

Location: 3 miles southeast of Cripple Creek

Two blocks of false-fronted buildings graced the main street of Anaconda. Hotels, saloons, and mercantile stores were intermingled with the shops and offices of an optician, doctor, printer and dressmaker. Surrounding the town were many mines.

The Mary McKinney, located on the hill east of town, was a major producer yielding over $11,000,000 worth of high-grade ore. Other good producers were the Anaconda Tunnel and the Jackpot.

The town, which was founded in 1894, had a peak population of well over one thousand persons at the turn of the century. The Florence and Cripple Creek Railroad snaked through Squaw Gulch and the heavy concentration of buildings in Anaconda.

A young church organist, Mary Louise Guinan, later became (Texas "Hello, Sucker" Guinan) of New York speakeasy fame.

During the winter of 1904, a fire, which started in a meat market, devastated Anaconda. No effort was made to rebuild the town.

TELLER COUNTY 121

Hot soup was ten cents at this early Cripple Creek restaurant. *Colorado Historical Society.*

Early Cripple Creek was of wood frame construction. *Colorado Historical Society.*

Cripple Creek's first church was constructed of canvas over wood framing. *Colorado Historical Society.*

Cripple Creek had become a substantial city when Harry Buckwalter took this photo June 14, 1899. Mount Pisgah rises on the left. *Colorado Historical Society.*

Victor townspeople carry their belongings into the street as the big fire approaches, August 21, 1899. *Denver Public Library, Western History Department.*

Following the fire of 1899 the Gold Coin Mill (left center) and much of Victor was rebuilt with brick as this photo taken one year later indicates. *Denver Public Library, Western History Department.*

Goldfield during its "boom" days. The population of 5,000 dwindled to 50 by World War II. *Colorado Historical Society.*

Gillett was christened "The City of Destiny." *Colorado Historical Society.*

Independence about 1900. The Hull City mining properties are in the foreground. *Colorado Historical Society.*

A Sunday crowd poses at the entrance to Pinnacle Park at Cameron. *Archives, University of Colorado at Boulder.*

Altman boasted that it was the highest incorporated town in the world (elevation 11,146 feet). *Colorado Historical Society.*

Overlooking Anaconda, with Raven Hill and the Mary McKinney Mine in the background. *Colorado Historical Society.*

SOUTH CENTRAL REGION

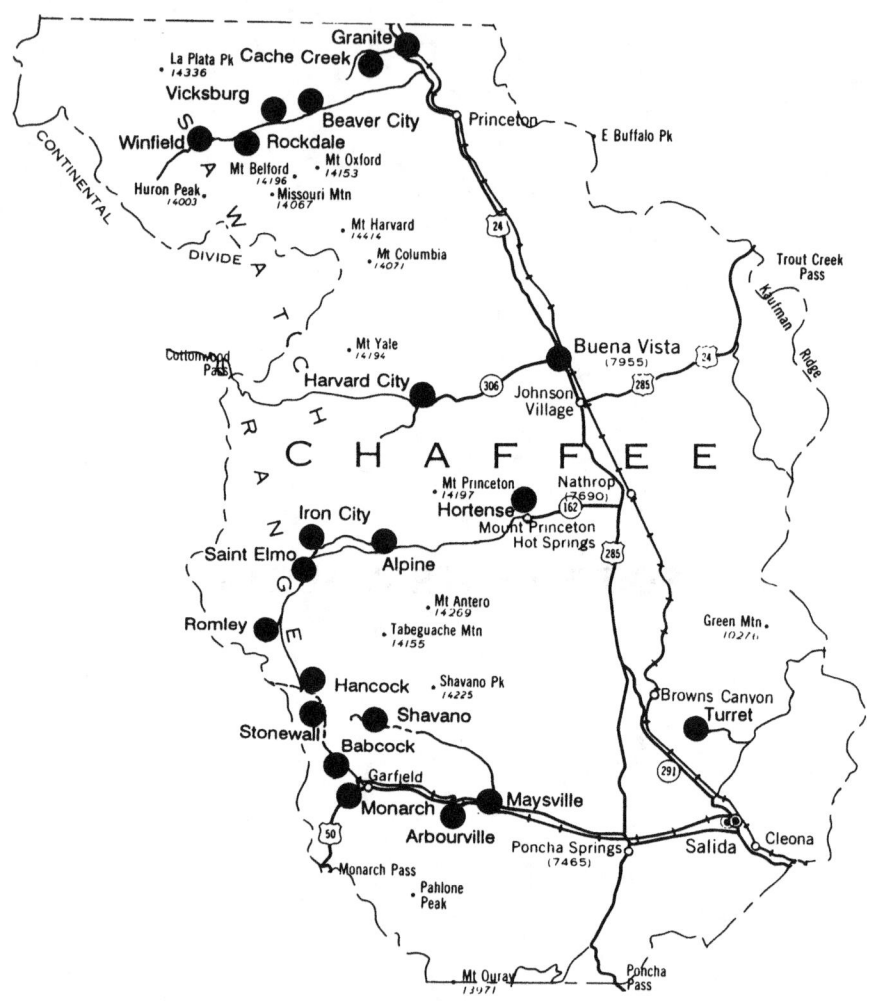

CHAFFEE COUNTY

Nuggets "big as eggs" were found somewhere west of Cache Creek in or near Lost Canyon, in 1860. The discovery became known as the Lost Canyon Placers. Not only was the canyon lost (at the time) but so were the prospectors — who could never find the location again. No one else has been able to, either.

CACHE CREEK

Location: 3 miles west of Granite

Placer gold was discovered near Cache Creek in the spring of 1860. The find was the earliest in Chaffee County. The settlement of Cache Creek was located in an arroyo on a high mountain plateau. It was a small but lively place. Although Cache Creek only had a population of about 200 people, it was a supply center for a cluster of even smaller settlements scattered through the hills to the west, such as Gold Run, Ritchie's Patch, and Oregon Creek.

The town, which was incorporated in 1866, had the first post office in Chaffee County. It also had some notable stop-overs. H. A. W. Tabor spent some time at the camp prior to his Leadville days. He learned some valuable lessons in mining while his wife learned some as a storekeeper. Another miner who spent some time at Cache Creek was S. B. Kellogg, who gained prominence during the early exploratory expeditions into the San Juans.

By 1872 Cache Creek and the surrounding area were all owned by a single mining syndicate — the Cache Creek Mining Company. It was then purchased by the Twin Lakes Hydraulic Gold Mining Syndicate. A water works of sorts was established in 1884 to redirect additional water to the vicinity from both Clear Creek and Cache Creek. This ingenious effort put old flumes and sluices back to work, and the gold output was nearly tripled.

Eventually the placers played out, and Cache Creek's residents moved on. The cemetery, gravel piles, and some foundations are all that remains.

"The rule of a gunfight is that the loser is always wrong."

GRANITE

Location: 18 miles south of Leadville on U.S. Highway 24

Shortly after some of the first gold discoveries in Colorado, placer claims were staked along both the Arkansas River and Cache Creek. This activity generated the birth of three mining camps in 1860 — Cache Creek, Granite and Georgia Bar. Although there was some successful placer mining at Georgia Bar, it quickly became insignificant. The other two became active camps. Longevity belonged to Granite which is still inhabited today. The town was laid out at the mouth of Cache Creek.

Placer mining was the most active form, and continued until about 1889 when hydraulic mining took over. Meanwhile in 1867 a couple of good lodes were located — the Yankee Blade and Belle of Granite. Both had a short life. The Yankee Blade, which was the better producer, yielded about $60,000. When hydraulic mining was discontinued in 1911, Granite was lulled to sleep. The town, where H. A. W. Tabor was once a storekeeper, had a population of about 500 during it's prime in the 1860s. Today, Granite is a quiet little community — but it wasn't always that way.

In 1875 when Granite was still the seat of Chaffee County, two scandalous and related murders occurred. Elijah Gibbs killed another Granite resident and quickly went to trial. The presiding judge was Elias Dyer, son of the famed itinerant Methodist preacher Father John Dyer. When Judge Dyer acquitted Gibbs, friends of the dead man immediately formed a vigilante committee and threatened revenge. After warrants were issued for the arrest of some of the vigilantes, Judge Dyer was shot to death. No arrests were made, and the matter became history.

HORTENSE

Location: 9 miles south of Buena Vista

The Hortense Mine, a rich silver producer, was discovered in 1871 by Captain Merriam. The mine was located on Mount Princeton, 12,000 feet in elevation, and was difficult to reach and work. A very narrow and dangerous trail wound up the mountainside. Later a road was built and oxen teams replaced pack trains to haul the ore off the mountain.

Hortense was almost "across the street" from Haywood Hot Springs (now Mount Princeton Hot Springs) where the famous Anteso Hotel was under construction during most of Hortense's existence. Ground was broken in 1877 but it took forty years to complete. The Anteso opened in 1917. It no longer exists.

The miners' cabins at Hortense blended in with those at Mount Princeton Hot Springs. It was hard to tell where one community stopped and the other started — as if anybody cared.

Hortense received a post office in 1879 and the railroad came in 1881. The population in 1887 was 100.

Eventually the ore gave out, the mine closed, and the miners vacated their cabins.

HARVARD CITY

Location: 6 miles west of Buena Vista via State Highway 306

Shortly after the discovery of gold along Middle Cottonwood Creek, the mining camp of Harvard City was established in 1874. The town, which preceded Buena Vista by some five years, was laid out where the South Cottonwood and Middle Cottonwood conflow to become Cottonwood Creek.

The Cottonwood Pass Toll Road crossed the Continental Divide west of Harvard City, followed Middle Cottonwood Creek to the townsite, then continued on to the future site of Buena Vista. For a few short years, Harvard City was an active supply center and stopping point for freight wagons, prospectors, and tourists. There was a dance hall and saloon.

Cottonwood Pass was a treacherous route — especially during winters. Snow slides and rock slides were commonplace. Several lives were lost along the trail. On one occasion, travelers had to tunnel 90 feet through the snow, then lower their wagons 700 feet down the mountainside in order to continue on.

Once the new road over Independence Pass was completed in 1881, the difficult Cottonwood Pass route fell more and more into disuse. Miners packed up and moved to the lucrative silver camps of Leadville and Aspen. Buena Vista emerged as the main supply center in the vicinity — and the community of Harvard City was all but forgotten.

BABCOCK

Location: 4 miles northwest of Garfield on the Chalk Creek Pass Trail

Of the mining camps that blossomed in the Mt. Aetna area, Babcock was the largest. The camp was established at the location of a rich silver strike in the late 1870s. The life of the camp was short, however, due to the difficulty of transporting ore from the isolated location high on the Chalk Creek Pass Trail.

The camp probably derived its name from Colonel Babcock who operated the Mountain Chief Mine at nearby Cree's Camp.

Being the only camp of consequence south of the pass, Babcock was a stop for travelers from Monarch or Garfield to Hancock and points north.

All that is left at the remote site today is one cabin with its walls still standing and several others two or three logs tall.

"Most hotels in the mountains have hot and cold water. Hot in the summer and cold in the winter."

MONARCH

Location: 3 miles southwest of Garfield on U. S. Highway 50

According to history, the discovery of the Madonna, Monarch and Little Charm mines are sometimes attributed to Nicholas C. Creede (of later fame, founder of Creede). Most believe that he discovered the Madonna — others say, the Monarch and Little Charm. At any rate, they, the Uncle Sam, and Columbus were all located in 1878. The Eclipse, Silent Friend, Smith, and Gray mines were discovered later. All were good producers. Monarch boasted of having over twenty productive mines within a half mile of town. With three hundred employees and three shifts operating, the Madonna shipped up to thirty car loads per day and was the top mine.

The tent colony that blossomed in the high mountain valley was originally called Camp Monarch. The date attributed to the commencement of construction is May 15, 1879. Almost immediately the name was changed to Chaffee City in honor of Senator Jerome B. Chaffee. In 1884 it was renamed Monarch.

Monarch seemed to have a little of everything which included the usual hotels, boardinghouses, plenty of stores, and saloons. Meals were served at Katie Finn's Hotel, the Saddle Fork, and the Welcome House. Other establishments in the business district had names such as Eureka Hall, Frank Ozman's Gambling Casino, and Bill Goord's Palace of Pleasure — where miners could find whatever. Monarch Park was a gathering place on Sunday afternoons. Festivities and sporting events were regularly held, including baseball games and concerts — featuring the Scenic Line Band, a local group.

Monarch peaked between two censuses, so population estimates vary greatly. Some say the population never exceeded that of Maysville (which peaked at about 1,000). Other estimates range to as high as 3,000 people. It is possible that those included such satellite camps as Hartville. Based on photographs, number of buildings, mining activity, etc., it is this writer's opinion that Monarch was larger than Maysville but that it could never have housed 3,000 at its peak — 1,500 to 2,000 seems more realistic.

Monarch's mines predominantly produced silver. In 1893, when repeal of the Sherman Silver Purchase Act created devaluation and panic, the town faded quickly as did most silver camps. Monarch has vanished. Only the cemetery and a few mining ruins remain in the valley.

ARBOURVILLE
Location: 4 miles east of Garfield on U.S. Highway 50

The remaining ruins of Arbourville are located in a heavily wooded area, adjacent to a fenced pasture, and on private property.

Arbourville which was never very large, was established in 1879. *Crofutt's Grip-Sack Guide of Colorado* stated that Arbourville's principal occupations were mining and stock raising. Actually, the town had a smelter, but the nearest mines of any consequence were around Garfield, three miles above.

The town was an important stagecoach stop on the road to and from Monarch Pass. There was a hotel, boardinghouse, and general store, but most important to the economy of Arbourville was it's brothel. It was the only parlor house in the vicinity and attracted men from near and far. It also brought more money into the community than any other business.

Long after the other residents of Arbourville moved on, Frank Gimlett remained. Known as the hermit of Arbour-Villa (as he called it), Gimlett spent his idle time writing. *Over Trails of Yesterday* was produced in nine short volumes which sold for 25 cents each. He wrote letters to President Franklin Roosevelt — and once got an answer. Gimlett, whose favorite movie actress was Ginger Rogers, attempted unsuccessfully to rename two mountains which reminded him of a part of her anatomy — as "Ginger Peaks." He once sent the government a bill for $50,000 for guarding the snow and ice on the mountains. He claimed that not one single shovel full had been stolen while he was protecting it.

SHAVANO
Location: 10 miles northwest of Maysville on the North Fork Valley Trail

High on the North Fork of the South Arkansas River above the mouth of Hunkydory Gulch, is the site of Shavano. When gold fever struck the Monarch Mining District, prospectors combed the mountains. A strike was made in 1879 and the camp was established. Lots twenty-five feet wide were given to anyone who was willing to grade and maintain that portion of the main street from its center line to the lot property lines. Many log cabins were constructed, as was a three story mill and a general merchandise store and saloon. Although there were future strikes in the area, Shavano was virtually dead by 1882, three short years after it began.

Mount Shavano, northeast of the mining camp, is the site of the "Angel of Shavano", subject of several legends told in different ways. One of the popular versions tells of a Ute Indian princess who had a great love for her people and the beautiful countryside. Long before the white man roamed these hills, there was a long extended drought which was driving the Utes from their land. Each day the Indian princess knelt near the foot of the Mount Shavano to pray for rain. Believing that she had been beckoned by the gods — that the drought would end and her tribe could live — she sacrificed herself. Upon her death she became the "Angel of Shavano". She appears each year to shed her tears on the fertile land below. There has never been another drought.

Alpine had no church, but one of the ladies' opened a Sunday School. When a child was asked what Christ was doing on the Mount, his reply was, "Guess He was prospecting."

ALPINE

Location: 16 miles southwest of Buena Vista via U. S. Highway 285 and State Highway 162

Originally Alpine had a short life and then turned into a ghost town. Later it was revived into the lovely community of summer cottages it is today.

The first house was constructed in 1877. Others followed and Alpine grew rapidly. It was a mining town, a supply town, a stopping point for travelers, and finally a railroad construction town.

The boom years were 1879, '80 and '81. When you mix miners with railroad construction workers the result is a rowdy town. In fact, it was rowdy before the arrival of the Denver and South Park Railroad in 1881 — there were already 23 saloons.

According to the 1880 census, 503 people resided in Alpine. Locals say it was at least twice that.

Col. Chapman, Alpine's first mayor, built a smelter which was overhauled and made into a sampling and concentration works. It employed forty men in 1883 and had the capacity to work thirty tons of ore per day.

The demise of Alpine began when the railroad reached St. Elmo. People started to pack up and move. St. Elmo showed more promise as a city and was closer to the most productive mines. The editor of Alpine's newspaper *The True Fissure* closed it and started a paper in St. Elmo. Alpine became more and more deserted.

ROMLEY

Location: 3 miles south of Saint Elmo

The remains of a cabin, which was wallpapered with 1913 newspapers, sits a mile above Romley near the Mary Murphy Mine. Dr. A. E. Wright and John

Royal located the mine in the mid 1870s. After they sold it in 1880, the Mary Murphy reaped a bonanza. It was far and away the largest producer in the Alpine district and largely supported the towns of Romley, Hancock, and St. Elmo. The Mary Murphy, along with the Alie Bell, Flora Bell and other lesser mines, yielded an estimated sixty million dollars in gold, silver, and other minerals during their years of production.

Romley, once called Morley and also Murphy's Switch, was built on a little flat which is high on a hillside midway up the incline between St. Elmo and Hancock. The Denver, South Park and Pacific Railroad arrived in 1881. A post office was established in 1886.

Sparks from a train engine ignited a fire which destroyed most of Romley in 1908. When the town was rebuilt, its buildings were painted bright red. They stood until 1990 when bulldozers leveled the dilapidated remains. Although it had its own schoolhouse and several commercial buildings, the town was very reliant upon St. Elmo.

HANCOCK

Location: 6 miles south of Saint Elmo

Although it was named for the Hancock Placer, the first claim in the area, Hancock predominantly grew as a railroad construction town. The Alpine Tunnel, the first built through the Continental Divide, was a major construction project. Of the hundreds of construction workers on the tunnel, many lived in Hancock, which was located at the base of a steep grade below the eastern terminus (see also Alpine Station). The town was established during the summer of 1880.

Main Street, with its row of false fronted buildings, faced the tracks of the Denver, South Park and Pacific Railroad. There were stores, saloons, and boardinghouses. Hancock's population reached about 1,000 at its peak.

The railroad's steep angle of descent from the tunnel entrance down to Hancock was said to be a harrowing trip or — as one of the passengers, Mark Twain, described it — "breathtaking."

A decline in the areas mining activity, and the demise of the Alpine Tunnel, spelled doom for Hancock. A snowslide in 1902 leveled many of the cabins. The last building, a saloon, collapsed a few years ago. Not much remains except the picturesque meadow where Hancock once stood.

STONEWALL

Location: 1 mile south of Hancock

Where Hancock stopped and Stonewall began we really don't know for there were once cabins scattered along the mile long road from one settlement to the other. Also a few cabins once lay in the high meadow near the mine.

The Stonewall Mine was the largest producer in the Hancock-Stonewall area and largest south of the mines at Romley. It was located in 1879 and was worked until 1915.

Stonewall had one store which was also the only saloon. The camp was very reliant on the commercial activities in Hancock for its existence.

When the mine shut down so did the camp. It was totally deserted over ten years before Hancock was.

SAINT ELMO

Location: 20 miles southwest of Buena Vista via U. S. Highway 285 and State Highway 162

St Elmo is one of the best preserved ghost towns in Colorado. Most of the buildings up and down Main Street look much like they did a century ago.

The town came into existence during the late 1870s, and was incorporated in 1880 under the name Forest City. When the post office wouldn't approve the name (because there was another Forest City, in California), it was changed to St. Elmo. The community, whose population peaked at approximately 2,000, prospered thoughout the eighties and nineties.

The Mary Murphy Mine was important to the life of the community and its tremendous success helped sustain the whole district. But there were several other mines in the vicinity, and St. Elmo had other things going for it as well. It became the major supply center within the district. Also, it was a convenient stopover for travelers to and from the several passes nearby — especially Tin Cup and Alpine. It was also a stage stop prior to the railroad's arrival in 1881. Once the last leg of the Denver, South Park and Pacific Railroad was complete and the Alpine Tunnel was opened, St. Elmo became a main station on the line.

Nearby Romley and Hancock had their saloons, but the place to be on Saturday night was St. Elmo. Never known as a wild town, it was colorful and people knew how to celebrate.

In addition to its saloons and dance halls St. Elmo had several stores and restaurants, five hotels — the Clifton and Pacific House being the most notable, two saw mills, a newspaper — the *St. Elmo Mountaineer* — but no church.

The wandering men of the cloth, Father Dyer and Bishop Machebeuf visited periodically in the early years and preached wherever they could get a group together. After the schoolhouse was constructed in 1882, church services and Sunday school classes were held there.

In 1890, fire destroyed two blocks of businesses. Fred Brush, who operated the drugstore where the original post office was located, left such "important" items as liquor and cigars, and suffered a burn in order to save the mail. He was an instant hero. The post office moved into the Stark Brothers' Store. Tony

Stark, his brother Roy who died early, and their sister Annabelle who grew up as a child in St. Elmo, operated a general store from the boom years until there was nobody left to buy anything. Tony and Annabelle saw the bud and the bloom, then watched it wilt. They were St. Elmo's last residents, hanging on for many years after everyone else had departed.

"A string around your finger helps you remember. A rope around your neck helps you forget."

BUENA VISTA
Location: 33 miles south of Leadville on U.S. Highway 24

In the late-afternoon shadow of three of Colorado's highest mountains lies the town of Buena Vista. At first the settlement was called Cottonwood, then Mahonville (for the Mahon family, early settlers in the area), and again changed to its current name. It is generally believed that the town was named by Alsina Dearheimer, the "Mother of Buena Vista." It means beautiful view in Spanish. The town was platted by W. M. Kasson who had acquired the land.

During the '70s prospectors combed the banks of the Arkansas River and Cottonwood Creek. Gold was discovered just over a mile south of Buena Vista and a camp established called Free Gold. A store, saloon and stamp mill were constructed. The camp which was operated by the Free Gold Company, had a population of nearly 150. This was about the only mining in the immediate vicinity of Buena Vista.

Buena Vista was established in October 1879 and became an important supply and shipping center for the mining industry. It was also a smelter town. With a silver boom to the north at Leadville, and a gold boom to the west at Saint Elmo, the streets of Buena Vista were full with clattering wagons loaded with ore, stagecoaches and carriages, and covered wagons.

Then came the Denver & Rio Grande Railroad. Trains could arrive or depart in most any direction. Rails across Poncha Pass and Marshall Pass to the south connected with the newly incorporated towns of Salida and Poncha Springs, then ran north to Buena Vista. Rails from the north followed the Arkansas River from Leadville. The Denver, South Park & Pacific Railroad crossed the Continental Divide by way of the Alpine Tunnel, down Chalk Creek through Saint Elmo, and into Buena Vista from the west. From South Park, tracks climbed over Trout Creek Pass and arrived into town from the east.

Buena Vista was voted as new county seat in the election of 1880. Salida appealed, charging Buena Vista with illegalities. The State Supreme Court upheld Salida's appeal and declared the election illegal — but awarded the county seat to Buena Vista anyway. County records were moved from Granite, and in 1882 a new county courthouse was constructed.

Buena Vista was a wild place in its early days. It is said that there were 68 places where a man could get a drink — such as the Mule Skinner's Retreat. Plentiful also were the brothels — the most notorious being the famed Palace Manor, with its madame, Elizabeth "Cockeyed Liz" Spurgeon. There were so many gamblers and prostitutes using assumed names that the post office had to establish a separate mail box for aliases. Early Buena Vista was virtually lawless. Shootings and killings were commonplace, with the offender often unpunished. In one instance a killer pleaded guilty, then paid a $10 fine. With the arrival of the famous "hanging judge," ironically named Judge Lynch, things changed. "Swinging" bodies were a common sight.

Buena Vista today is a quiet community. It is predominately a ranching and farming area, and site of the state reformatory. Many historic buildings remind one of the town's history. The old courthouse now houses the Buena Vista Heritage Museum. The first church, which was constructed in 1880, is presently the Chamber of Commerce. The building that once housed the Palace Manor of "Cockeyed Liz" still stands — as do many others. President Harry S. Truman once had a summer residence in Buena Vista.

"He never learned to spell 'cause the teacher kept changin' the words."

MAYSVILLE

Location: 6 miles east of Garfield on U. S. Highway 50

When the Monarch Mining District boomed during 1879-80 several towns sprang up between Poncha Springs and Monarch Pass — Maysville, Arbourville, Garfield, and Monarch. And they did so simultaneously. Maysville was platted on property of the Feathers Ranch owned by Amasa Feathers, who was at the right place, at the right time and took advantage of it. Miners swarmed in, purchased lots for up to $75 each and erected tents. The community was a busy place in just a matter of days. It was named for the Kentucky hometown of General William Marshall, who blazed the trail over Marshall Pass.

Two smelting works were constructed at Maysville which treated much ore from the surrounding mountains. The location of the smelters was ideal. Toll roads descended into Maysville — one from Monarch Pass, another from Shavano. Arrival of the railroad in 1881 further facilitated shipping.

A fire in July 1880 destroyed five buildings located in the business district. They were rapidly rebuilt and the economy of Maysville didn't suffer. During its prime, about 1882, Maysville had a population of 1,000, several hotels, a bank, schoolhouse, numerous shops and saloons, a lumber yard, the aforementioned smelters, and two newspapers. The *Maysville Chronicle* boasted of having the largest circulation in the county. The *South Arkansas Miner* was printed in folio form.

Through the late '80s area mining declined. Mining companies still producing seemed to find better smelters elsewhere. Maysville was experiencing hard times well before the silver crash in 1893. By then the population was minimal.

"The only place some folks make a name for themselves is on their tombstones."

IRON CITY

Location: *20 miles southwest of Buena Vista, near St. Elmo. Turn right at the entrance to St. Elmo*

Iron City was a small community nestled in the trees near St. Elmo, and below the once existent Iron City Reservoir. It was short-lived.

A smelter was constructed at the townsite in 1880 which treated much of the ore from surrounding mines. An electric generating plant also was built which furnished power for a dredge at Tin Cup. Power poles ran from Iron City up the mountain, over Tin Cup Pass, and on to the dredge site.

The railroad bypassed Iron City enroute to St. Elmo. Once completed, ores were shipped by train, and both Iron City and its smelter lost their usefulness.

Eventually the reservoir, and what was left of the city, washed away.

The Iron City cemetery, with its gravesites scattered through the trees, can be seen about a mile down the road that turns right just east of the entrance to St. Elmo.

BEAVER CITY

Location: *9 miles west of Granite via U.S. Highway 24 S. and Forest Route 390*

When prospectors found good float near the site of Beaver City, they immediately established a camp. Because of its proximity to Granite, Beaver City flourished for a short time. It was mainly a tent community with very few permanent buildings constructed. As better producing mines were discovered further up the gulch, and as more substantial towns were being built, such as Winfield and Vicksburg, the residents of Beaver City packed up and moved. A couple of cabins remain at Beaver City.

VICKSBURG

Location: 11 miles west of Granite; west of Clear Creek Reservoir and U.S. Highway 24 on Forest Route 390

Nestled in an aspen grove in Cochetopa National Forest, lies the remains of one of the most rustic mining towns in Colorado. A dozen well-kept and remodeled cabins adorn one side of picturesque Main Street which is lined on each side with a perfectly straight row of trees. Opposite the cabins are two small museums.

Residents of Vicksburg worked the many mines scattered throughout Clear Creek Gulch. As an active mining town, the life of Vicksburg was short lived. The post office was established May 1881 and discontinued July 1885. During that time, the population peaked at 250 persons.

ROCKDALE

Location: 13 miles west of Granite - on Forest Route 390, between Vicksburg and Winfield

There were nearly two hundred mine locations staked or worked in the gulch surrounding Clear Creek. Because the mining operations were spread out, so were the mining camps. Rockdale is nearly equidistant between Winfield to the west and Vicksburg to the east, and is located on Clear Creek.

Rockdale flourished at the same time as its neighbors but was smaller. Today four of the original cabins remain — situated in a neat row at the heavily wooded site.

WINFIELD

Location: 15 miles west of Granite; 4 miles west of Vicksburg on Forest Route 390

Winfield sprang into existence in 1880 at the junction of the north fork and south fork of Clear Creek. Cognizant of the speed with which a fire can spread, Winfield was platted with a safe distance between buildings. At the time the post office was established in July 1881, there was a mercantile building, a boardinghouse, a schoolhouse in which church services were held on Sunday, and many scattered cabins.

Many Winfield residents worked the Banker Mine, the largest producer in the area. Several other mines aided the short prosperity of Winfield. In about 1882, the population of Winfield peaked at 250. The silver devaluation in 1893 caused most of the residents to pack up and move out. Some mining activity continued for several years. The post office was discontinued in September 1912.

Many buildings remain today. The schoolhouse, now a museum, was restored, as was the Ball Cabin across the street.

"Chasin' gold is like chasin' your shadow — the harder you chase it, the faster it gets away from you."

TURRET

Location: 12 miles north of Salida in Cat Gulch south of Green Mountain

The road into Turret transverses the main street upon which most of the buildings are located. One freshly painted white two-story house stands amid what is otherwise a ghost town — and an interesting one at that. About seventeen (unpainted) structures still stand in various states of dilapidation. On a small knoll overlooking the town lies the cemetery.

There was a two-story hotel complete with a balcony and wallpapered interior, post office, schoolhouse (which included living quarters for the teacher), general merchandise store, saloon, and a butcher shop. Twice each week a stagecoach arrived from Salida. A newspaper, *The Gold Belt*, was published for a short time around the turn of the century. The town which was platted in 1897 had no elected officials — only an appointed marshal.

It all started with the discovery of the Gold Bug and Vivandiere mines in 1896. Optimism ran high, but neither mine materialized very well. The Golden Wonder was another disappointing property. The Independence Mine was the largest producer and shipped ore until it was closed in the winter of 1916 by its owner, the Turret Copper Mining and Reduction Company. The Denver & Rio Grande Railroad built a spur to the camp and mine at Calumet in 1898. Ore from the Turret mines could be hauled by wagon about five miles to the railhead where it was transferred to railroad cars.

Although some mining was done for about twenty years, the real boom only lasted for about five years — 1898 through 1902. After that the population declined rapidly.

Granite about 1890. *Colorado Historical Society.*

Monarch, nestled in a high valley. *Colorado Historical Society.*

CHAFFEE COUNTY 143

Romley was also known as Murphy's Switch. All of the buildings in the flat (at right) were once painted red. *Denver Public Library, Western History Department.*

The depot at Alpine. *Denver Public Library, Western History Department.*

Buena Vista following completion of the Chaffee County Courthouse (at left). *Colorado Historical Society.*

St. Elmo's Main Street, (looking east.) *Collection of Dave Southworth..*

CHAFFEE COUNTY 145

The false fronts of Hancock faced the railroad. *Colorado Historical Society*.

The Gregory Hotel at Turret had a wallpapered interior.
Denver Public Library, Western Historical Society.

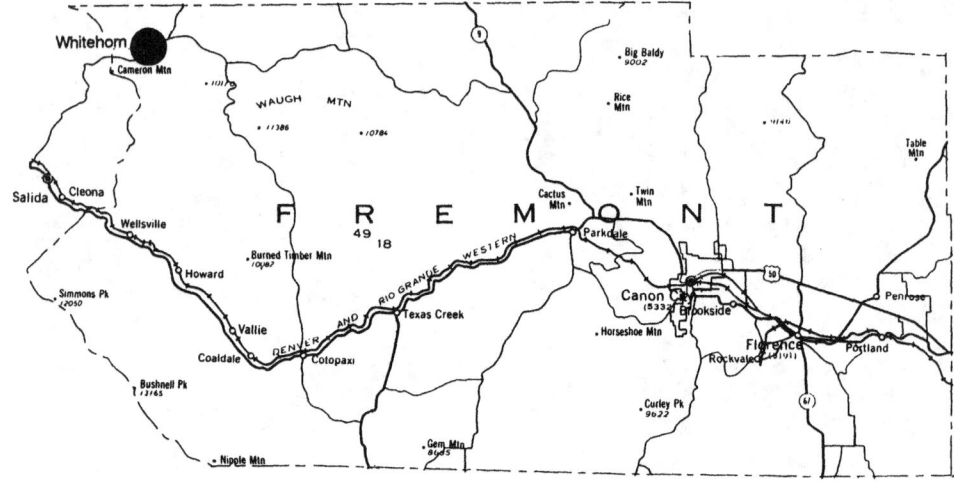

FREMONT COUNTY

An unidentified rich ore sample once assayed at $7,500 per ton. The assay office didn't know who the owner was because the shipping label had been lost. It may be that somebody could have struck it rich had they only known the results.

WHITEHORN

Location: 15 miles northeast of Salida via Forest Route 175

Whitehorn was a late-blooming gold camp. The first strike was made in 1897 and grubstaked by A. L. Whitehorn for whom the town is named. More solid discoveries followed immediately. The Cameron Mines Land and Tunnel Company purchased several claims, including Whitehorn's, and platted the town. Frank and Harry Woods of Gold Coin Mine fame (see Victor) bought up another 70 claims in the area. The Woods Investment Company also pumped much money into the community. Through the endeavors of the two companies, an aesthetically attractive town emerged — complete with sidewalks.

The mining camp flourished rapidly and by the end of 1898 had nearly 1,000 residents. A fine ten-room hotel was constructed — the Witting. A post office was built, as was a telephone company whose service connected with Salida. Main Street had a large saloon, cafe, several stores and a couple of brothels. A schoolhouse was erected on the edge of town. There was one newspaper, the *Whitehorn News*. The town was served by two stagecoach lines.

Whitehorn had one mill to service area mines. The nearest smelter was at Salida. Most notable of the area mines were the Independence, Golden Eagle, Cameron, Molly Gibson, Guess, and the Ida.

A fire began at the Witting Hotel the evening of May 24, 1902. The blaze swept down Main Street ravaging the surrounding buildings. Before the bucket brigade (there was no fire department) could control the flames, half of Whitehorn had been destroyed. Little was done to rebuild the town after the conflagration.

The remaining buildings of Whitehorn are on fenced private property. They are visible from the road but permission should be obtained before entering the property.

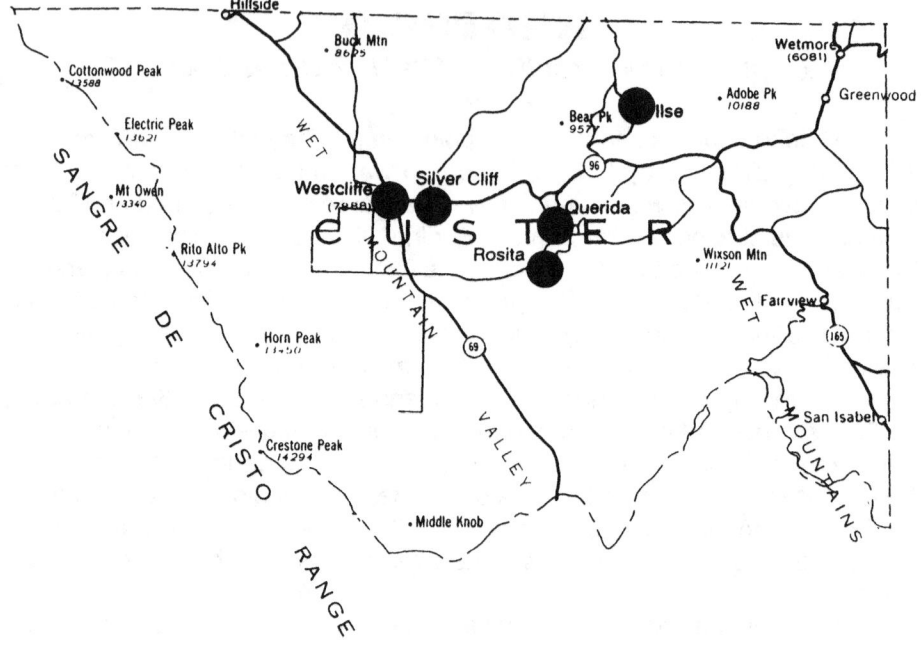

CUSTER COUNTY

The large cheese factory at Rosita produced nearly 250 pounds of fine cheese per day, until — the cattle discovered the delight of eating the wild garlic which grew in profusion. Their milk became contaminated, and the factory closed.

ROSITA

Location: 11 miles southeast of Westcliffe

In 1872 soon after the earliest claims were staked, the Hardscrabble Mining District was organized. The town of Rosita (originally named Brown's Spring) sprang into being the following year. It was the original Custer County seat and the principal community in Wet Mountain Valley until Silver Cliff experienced its boom in 1878.

Dick Irwin, for whom the town of Irwin was named (see Irwin), made one of the earliest strikes near Rosita. Two of the better producing mines were discovered in 1874 — the Pocahontas and the Humboldt. The Pocahontas was involved in two "claim wars." On one occasion an ex-con named Graham and his group of toughs were hired by some of the "claimants" to protect their interests. During an ensuing shoot-out in the center of town, Graham was killed and the others run out of town.

Rosita had its share of violence. In another ill-fated incident, two drunks "crashed" an Odd Fellows dance. They were quickly "bounced" into the street. Shortly thereafter, the drunks returned, armed — then shot and killed the "bouncers." The community was incensed. According to legend, printed invitations were circulated and the whole town attended the lynching.

The community of Rosita (Spanish for "little rose") grew rapidly during 1874-5. The Grand View was the finest of several hotels. There were many stores, a bank, two newspapers, and most notably a large brewery. Rosita was one of the earliest towns in Colorado to have telephone service. Additionally there were two stamp mills and two reduction works. The population topped 2,000 at its peak.

A devastating fire swept through town on March 10, 1881. As the population had dropped considerably (1,008 according to the 1880 census) prior to the conflagration, Rosita was never fully rebuilt.

Rosita received one more face lifting, however, in 1957. Metro-Goldwyn-Mayer partially refurbished the town for a movie set. Buildings were moved, some roofs repaired, and a few false fronts and hitching rails were added. "New" faded signs were installed on buildings for identification. The site was used for part of the filming of "Saddle the Wind," starring Robert Taylor.

QUERIDA
Location: 9 miles east of Westcliffe

E. C. Bassick, a miner, passed through Wet Mountain Valley each day on his way to and from his job on Tyndall Mountain. While traveling his route one day in 1877, he discovered a rich deposit. The ore was a composite of many minerals including galena, gold and silver. He staked a claim, named it for himself, and quit his job. For a time, the ore was so rich that it was shipped to Westcliffe under armed guard. Bassick extracted a half million dollars in gold and silver in less than two years. He sold the Bassick Mine to an eastern syndicate in 1879. The mine continued to produce and eventually had a yield of several million dollars.

The settlement which grew up near the mine was originally named Bassickville. Later it was changed to Querida, which is spanish for "beloved". The town had a smelter, sawmill, hotel, and a few stores. Querida peaked about 1882 with a population of 500.

Litigation shut down the Bassick Mine in 1885 and the town's population began to dwindle. There was a rebirth when the mine opened once again. In 1903 the Melrose Gold Mining Co. bought up and worked many claims in the area. Querida, however, was becoming a ghost town. The post office which was originally established in 1880 was discontinued in 1906.

"The best way to learn how much money is worth — is to try to borrow some."

ILSE
Location: 28 miles southwest of Florence

When stories are told and retold, distortion may appear in the form of exaggeration or change in the name of places and persons. Such appears to be the case in the story of the Ilse immigrant who had a "change of heart." According to one story Joseph Raphael De Lamar spent all of his time and money attempting to develop the Terrible Mine (Ilse), went broke, and left Ilse without paying his creditors. Another story indicates that Frank Andracich, after finding a rich deposit while farming, spent all of his time and money attempting to locate the mother lode, went broke, and left Ilse without paying his creditors. The immigrant (whoever he was) struck it rich in Idaho several years later. His change of heart brought him back to the Wet Mountain Valley where he sought out his creditors and paid off all his debts. [Note: There is no county record of Frank Andracich, which doesn't mean he didn't live here. J. R. De Lamar paid property taxes in 1880.]

Never a large community, Ilse's population fluctuated with activity at the Terrible, known by locals as "the mine" (another name for the same property is the Ilse Lode). From 1878 to 1888 the mine produced well, and the town had a few hundred residents. There was a hotel, boardinghouse, general store, and three saloons.

Fire ravaged much of Ilse in 1887. Mining production was dwindling at the same time, so very few of the fire-damaged structures were replaced. Some mining continued after the turn of the century, the greatest activity being in 1903 when a new mill was constructed.

According to legend, a cave full of Spanish treasure exists somewhere in the Sangre de Cristo Range. A mysterious cavern — Caverna del Oro — is located high on Marble Mountain. Many Spanish artifacts have been found at the site, but no treasure. Is is possible that the legendary lost treasure might be hidden somewhere nearby?

SILVER CLIFF

Location: 1 mile east of Westcliffe on State Highway 96

Silver was discovered in the vicinity during the summer of 1877 by R. J. Edwards. Although the content was low-grade, it was sufficient enough for Edwards to continue his search. Within one year he hit the jackpot with silver that assayed up to 740 ounces per ton. Prospectors flocked in — claims dotted the area — and the boom was on. Edwards' Racine Boy Mine, and others, yielded over 2,000,000 dollars during the first year.

Silver Cliff was platted with exceptionally wide streets. Within weeks many commercial buildings and over 200 dwellings had been erected. With twenty-five saloons, several gambling dens, and many dance halls Silver Cliff was a loud and rowdy town. There were several newspapers, two banks, stores of all types, and ten hotels (three were excellent — the others cheap, it is said). Additionally, there were four stamp mills and two reduction and concentration works.

As miners brought their wives and families to Silver Cliff a little culture was established. A school was built — and a couple of churches. By 1880 the population exceeded 5,000.

Through a county-wide election Silver Cliff became the county seat. Rosita, however, wouldn't part with all the county records. So, they were conveniently "moved" during a midnight raid on Rosita by the citizens of Silver Cliff.

Despite the fact that Silver Cliff had an award-winning volunteer fire department, the community was devastated by two fires.

After four boom years, several factors aided the decline of Silver Cliff. Part of the business district was never rebuilt after the second fire. Arrival of the Denver & Rio Grande Railroad into nearby Westcliffe lessened the importance of Silver Cliff. The final blow was the devaluation of silver created by repeal of the Sherman Silver Purchase Act. Much of Silver Cliff remains as a remembrance of those years of splendor. Memorabilia from the era can be seen at the old firehouse and town hall which is now a museum.

Having not heard from his prospecting brother George, Bill Skinner set out to find him. George had struck gold on Horn Peak which was rich enough to "blind a body." When Bill arrived all he found were the bodies of George and his burro who had killed themselves in a fall. Bill Skinner buried their remains but never found the rich mine.

WESTCLIFFE

Location: 52 miles west of Pueblo on State Highway 72

Clifton was once a tiny "suburb" of bustling Silver Cliff just one mile to the east. The Denver & Rio Grande Railroad was searching for a more southerly route in order to avoid the Sangre de Cristo Mountains as much as possible. Dr. J.W. Bell was instrumental in their decision to establish a station house at Clifton. After the tracks arrived in 1881, the community was renamed Westcliffe for Bell's birthplace, Westcliffe-on-the-Sea in England.

The beginning of Westcliffe was a direct result of the nearby silver boom, but it's growth was due to the railroad. Ores which once left Silver Cliff by wagon were later processed through Westcliffe and left via the Denver & Rio Grande.

As Silver Cliff faded, many residents moved to Westcliffe. Buildings were also moved — including the posh hotel, the Powell House. By 1928 the county seat had shifted from Silver Cliff to Westcliffe and the latter became the larger town by far.

Rosita before the fire of 1881. *Denver Public Library, Western History Department.*

Log buildings and tents at early Querida.
Denver Public Library, Western History Department.

Silver Cliff's Main Street in 1894. *Denver Public Library, Western History Department.*

Westcliffe, in the Wet Mountain Valley, as it looked in 1902.
Denver Public Library, Western Department.

SAGUACHE COUNTY

BONANZA

Location: 15 miles west of Villa Grove via Route LL56

Shortly after the second rush to Leadville, Bonanza and the other Kerber Creek camps were established in 1880. Sedgwick, Kerber City, Exchequerville, and Bonanza were close neighbors and friendly rivals. Bonanza (originally called Bonanza City) became the most prominent of the communities.

An ecstatic prospector yelled, "Boys, she's a bonanza!" — and the camp had a name. The Exchequer and Bonanza were the first two mines discovered in the vicinity. Other good strikes followed. The Empress Josephine yielded $7,000,000 during its earlier years. The Rawley, Wheel of Fortune, Eagle, and Defiance were also profitable properties. Mining activity has continued off and on to this day. Former President, Ulysses S. Grant once made offers to purchase both the Bonanza and Exchequer mines, but neither deal materialized.

Bonanza was a rowdy, bustling town. There were thirty-six saloons, seven dance halls, and parlor houses full of "sporting women." They say the population tripled on Saturday nights.

In addition to the "recreational facilities," Bonanza had several hotels, restaurants, stores of various specialties, a livery stable, and a newspaper — the *Bonanza News*. There were several mills and other mining properties as well.

Most of the ores were low-grade, making them more expensive to process. When silver prices took a dip in 1882, some of the population of 1,300 began to move on. Mark Biedell, one of Del Norte's leading merchants, sustained Bonanza for a while when he built a new concentration mill. Prior to the turn of the century, however, the population had dwindled to about 100. A couple of revivals recurred during the twentieth century, but neither amounted to much.

Author Anne Ellis, a long-time resident, wrote *The Life of an Ordinary Woman* in which she depicts life in Bonanza during the boom years and also the tough years.

LIBERTY

Location: 15 miles southeast of Crestone, east of the Luis Maria Baca Grant.

In 1822 King Ferdinand the Seventh of Spain bestowed upon one of his subjects the title of Don Luis Maria Cabeza de Vaca and granted him several hundred thousand acres of land in the west which included a 100,000 acre tract in Saguache County. The grant is known as the Luis Maria Baca Grant No. 4. The northeastern portion of the tract cuts into the Sangre de Cristo Range where deposits of gold, silver, lead, copper and iron were found. Several mining camps were established within the boundary of the grant — Duncan, Cottonwood, Lucky, Spanish, Teton and Julia City. Although there were small camps in the vicinity during the late 1870s, the greatest development occurred in 1890. Duncan and Cottonwood were the largest of the settlements, each with about 1,000 residents. When the value of land and minerals became evident, grant owners began proceedings to close the settlements and evict the miners. After the government upheld the provisions of the grant in 1898, the camps were closed in 1900 and miners were forced to move on.

The community of Duncan was located just inside the eastern boundary. When eviction occurred, miners moved their buildings and all a short distance outside the grant and named the new community Short Creek, then Liberty — so named because they were "free" of the grant and its problems.

William Gilpin (later to become Governor of Colorado) was among the early prospectors in the area. He discovered some gold while traveling with John Fremont, the famed pathfinder. He helped establish the original camp which became deserted prior to the 1890 boom.

Duncan had a schoolhouse — with 40 students, a few stores, a bunkhouse, and a newspaper — *The Golden Eagle*. Most everything was moved to Liberty. A few miners remained on the grant to work for the San Luis Valley Land and Mining Company which had taken over the mines inside the boundary. The schoolhouse and several cabins were later moved again — this time to Crestone — in 1910 when most of the mining was discontinued.

Once Liberty was established, more money was pumped into the vicinity. The Blanca Mutual Mining and Milling Company built a five-stamp mill in 1902. Mining dwindled during the ensuing years, however, and by 1909 the Liberty schoolhouse had but three students — all from the same family. A hermit maintained residence at the camp long after everyone else had left. He was "touched," it is said, and collected animal bones which he displayed for the occasional passer-by.

"Secondhand gold is as good as new."

CRESTONE

Location: 12 miles east of Moffat

When trapped by a storm, two prospectors found a cave in which to take refuge. Upon lighting a candle, they discovered three human skulls — and three bars of pure gold. They returned to Crestone with the gold, and certain there was more to be found set out for the hills once more. They searched and searched but could never find the cave again.

Nestled below the Crestone Needle in the Sangre de Cristo Mountains lies the townsite of Crestone. Prospectors began picking at the rocks in these mountains about 1879, and found some gold. A mild flurry of activity occurred and then tapered off. A post office was established in 1880 and Crestone was on the map. Free-milling gold was found in 1890 and a greater boom developed. It was during this time that Crestone became a substantial settlement. During the '90s the old Cleveland Mine was reopened, and the nearby camp of Wilcox sprang up north of town. Lesser camps were scattered through the foothills for miles.

By 1900 the population of Crestone had reached 2,000, and activity was at its peak. A railroad had been constructed from Moffat. There was a hotel, five general stores, several specialty shops and saloons, and prime commercial lots which sold for up to $600. The community was situated just beyond the northern boundary of the Luis Maria Baca Grant (see Liberty), and many residents lived within the perimeter of the grant. Many of the best mining properties were also within the grant's boundaries. In 1898 the Supreme Court upheld the provisions of the Luis Maria Baca Grant giving the owners excessive powers to create leases based on royalties. Meanwhile an eastern company purchased the mineral rights on the grant and a half interest in the townsite of Crestone. After some violence, the U.S. Circuit Court issued a summons to evict the remaining "squatters", allowing only those who were employees of the San Luis Valley Land and Mining Company to remain. Many of those who moved from the Baca Grant property re-established residence outside the boundary in Crestone.

Crestone settlers watched the mines they had worked so hard to build become the property of others. Some persevered however. Some new mining activity occurred in 1901 which bolstered hopes that the sagging economy might be sustained. The optimism was short-lived however, and Crestone settled back to become the quiet community that it is today. Many of the mines within the grant proper produced well into the twentieth century. Cattle raising is now the principal business on the huge ranch known as Luis Maria Baca Grant No. 4.

IRIS

Location: 10 miles southeast of Gunnison

Iris and her sister camp Chance were established the same year in 1894. Chance is just over a mile to the northwest, and across the Gunnison County line. The life of Iris, larger of the two camps, paralleled that of Chance. Both had a short boom — then faded quickly.

After Lehon and Turner initially discovered free gold, prospectors began combing Mineral Hill and the surrounding area. When the town sprang into being it was first called Union Hill, and then renamed Iris, presumably for all the wild iris growing in the vicinity. The Denver City Mine was virtually "in town," and one of the better producers. Other mines were the Alliance, Governor Waite, West Point, Tribune, Eureka, Gold Hill, Hidden Treasure, Friday, and Gold Sterling.

Iris had several shops and saloons, mail service three times per week, and telephone service from Gunnison. At its peak it is estimated that the combined population of the two camps exceeded 1,000.

Although expectations were high, and some decent ore was mined, overall production was not profitable. In 1897 three short years after its beginning, Iris faded rapidly.

The Canon House at Bonanza. *Colorado Historical Society.*

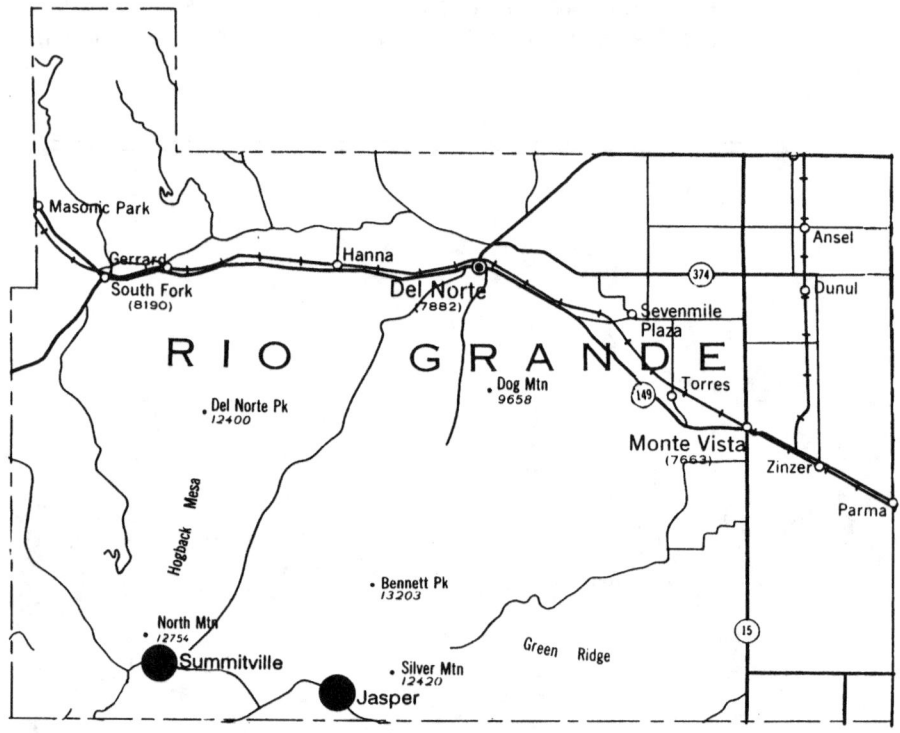

RIO GRANDE COUNTY

"A winning poker hand is like a gold strike — few and far between."

SUMMITVILLE
Location: 24 miles southwest of Del Norte

John Esmund discovered gold high on South Mountain in 1870. Each summer he worked the site, extracting an abundance of rich ore. He returned in the spring of 1873 to find that his claim had been jumped. Others were working his mine. Esmund had failed to do the necessary paperwork to properly establish his claim. The Little Annie became the best producer in the area. Although he was discouraged, Esmund knew there was much more gold in the area. He moved around the slope and made two new strikes — the Major and the Esmund (later called the Aztec). Having learned a valuable lesson, he did the proper paperwork on these claims.

Two brothers, James and William Wightman (for whom Wightman Creek is named), also discovered gold on South Mountain in 1870. Other strikes followed, and Summitville became a bustling gold camp.

By 1882 Summitville had a population of over 1,000, several hotels, a newspaper *The Summitville Nugget*, many stores and saloons, and nine mills to process ore from the mines.

The nearby city of Del Norte was important to the history of Summitville. Del Norte was an important shipping point, stagecoach junction, and supply center. It also had a resident who was instrumental in the success of Summitville. The suave and colorful Tom Bowen struck it rich with his Little Ida Mine. Bowen wore many hats. At one time or another he was a lawyer, Governor of the Idaho Territory, brigadier general in the Union army, district judge, Senator, and heavy gambler. The Little Ida was the catalyst which created a fortune for Tom Bowen. He purchased many other mining properties, including the Little Annie.

In 1885 several of the mines including the Little Annie ran into financial difficulties. The population of Summitville rapidly declined. By 1892 the town had only a handful of inhabitants. Mining in the vicinity experienced a few revivals. One occurred just before the turn of the century and another in the 1930s. The greatest production, however, has been in recent years. The heavy trucks and equipment of Galactic Resources, Ltd. have been rolling back and forth through the uninhabited and slowly deteriorating ghost town of Summitville. Galactic suffered heavy losses through 1991, sold the balance of its mining interests, and in late 1992 considered a reclamation project as a way to bounce back. The January 28, 1993 *Wall Street Journal* advised that Galactic Resources, Ltd. had filed for bankruptcy. Many buildings still exist at the high-mountain site of Summitville — one of Colorado's more interesting ghost towns.

"Too little temptation can lead to virtue."

JASPER

Location: 22 miles east of Summitville on Forest Route 250

In 1880 several strikes were made in the mountains around the Alamosa River. John Cornwall and a group of founding fathers established a settlement in the high mountain valley along the river.

The community was originally named Cornwall — then changed to Jasper. Nearby, Cornwall Mountain retained its name. When the post office was established in November 1882, Cornwall was named postmaster. One of the earliest settlers, Alva Adams, later became Governor of Colorado.

A fair amount of mining was done at the Sanger Mine, the Perry Mine, and others. In 1887 ten tons of ore were shipped to a smelter in Denver for refinement. Jasper miners never knew the result of the analysis because the smelter burned to the ground with the ore inside.

There must have been many tee-totalers amongst the miners in the area. Jasper's only saloon went out of business long before the town began to fade.

Following gold strikes on South Mountain, Summitville became a bustling mining town. *Denver Public Library, Western History Department.*

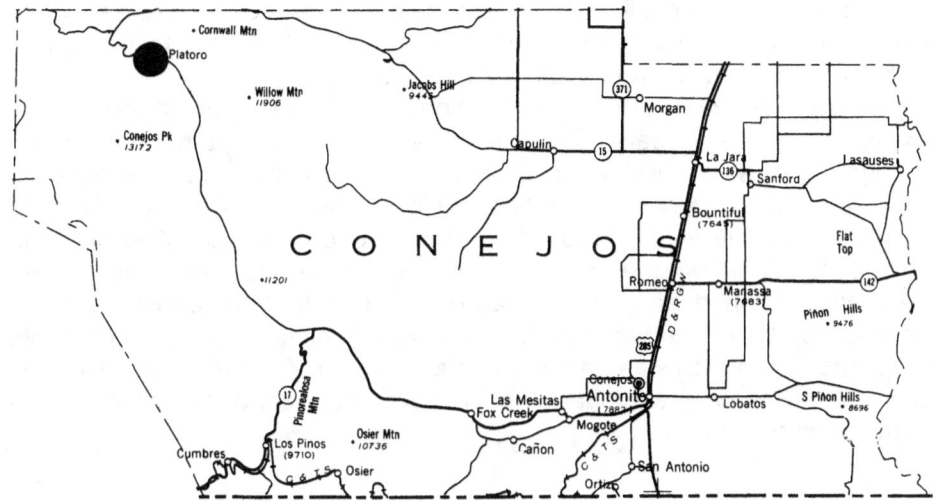

CONEJOS COUNTY

PLATORO

Location: 20 miles south of Summitville via Forest Route 330 and 380

Platoro sprang to life in 1882 when rich ore was discovered in the area. The town was platted in a flat, high mountain valley (altitude: 9,700 ft.) near the most northerly bend of the Conejos River.

Two mountains south of town bear the names of two of the earliest mine discoveries, the Mammoth and Forest King. Other good mines were the Anchor, Puzzler, Last Chance, Parole, and the Pass-Me-By.

Platoro was one of the highest mining towns in the region. It had a tough existence for several years. Ores had to be packed out by burro and supplies brought in the same way. A wagon road from Summitville was completed in 1888, and Platoro became more accessible. A post office was established in March of the same year. The community had a population of about 300 in 1890.

Ores taken from the area were never more than average in quality. As production declined many miners packed up and left. Platoro experienced a couple of up-turns after the turn of the century but neither developed into much.

The town which was named for "plata" and "oro", Spanish for silver and gold, is very much intact today. Cabins have been renovated and many are summer tourist rentals.

Otto Blake's large cabin is at left. The canvas structure beyond it is the Platoro Hotel. The general store flies an American flag, at right. *Denver Public Library, Western History Department.*

The Denver & Rio Grande train (at center) rolls through Russell in 1887. *Colorado Historical Society.*

166 SOUTH CENTRAL REGION

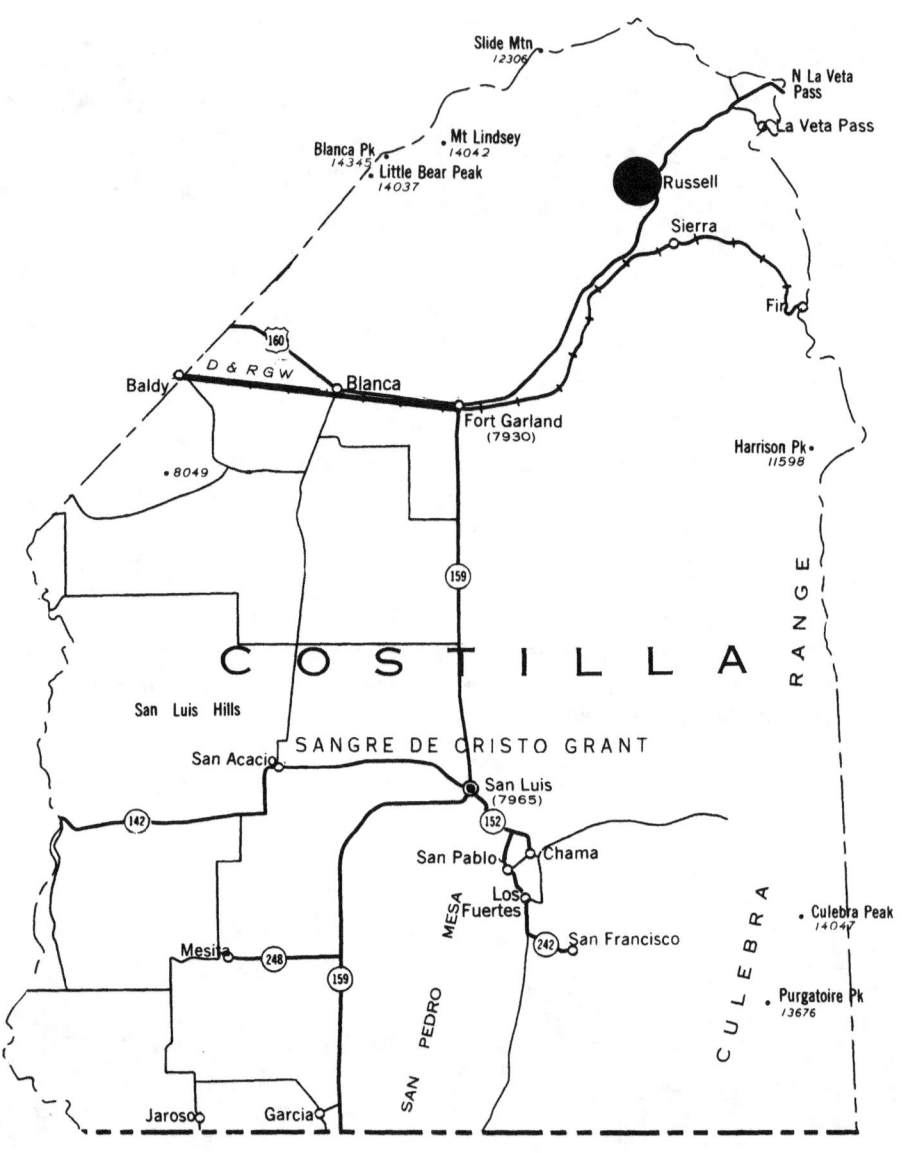

COSTILLA COUNTY

According to rumor, two Mexicans helped Juan Carlos bury $800,000 in gold near Blanca Peak west of Russell. The helpers mysteriously disappeared. Were they permanently silenced in order to keep the location a secret?

RUSSELL

Location: 14 miles northeast of Fort Garland via U.S. Highway 160

Below LaVeta Pass at the foot of the Sangre de Cristo Mountains placering activity on Sangre de Cristo Creek and Placer Creek brought settlers together in the 1870s and a camp was established. The settlement which blossomed at the confluence of the two creeks was originally called Sangre de Cristo.

Although gold, silver and copper were discovered and a large number of claims were staked, mining activity was limited. Several of the mines were consolidated and operated by the Colorado Coal and Iron Company for a short while. Many other claims remained undeveloped.

Eventually the settlement changed its name to Russell. The Denver & Rio Grande Railroad established a narrow gauge line through Russell and over LaVeta Pass. Its life was short, however. When the standard gauge line was built over the Sangre de Cristo range six miles south of LaVeta Pass, it bypassed Russell three miles to the south. The narrow gauge tracks were removed and the depot relocated. As a result, Russell became rather isolated and its population declined.

The community had a hotel which was operated by Margaret Sutton. Russell also had a sawmill, schoolhouse, post office, saloon and two general stores. The cemetery is said to have a few sluice box robbers among its gravesites. The population which reached nearly 300 at its peak had dwindled to only 65 by 1887. Later, the community was revitalized by the livestock industry. Sheep and cattle sustained the town for several years, then once again Russell faded.

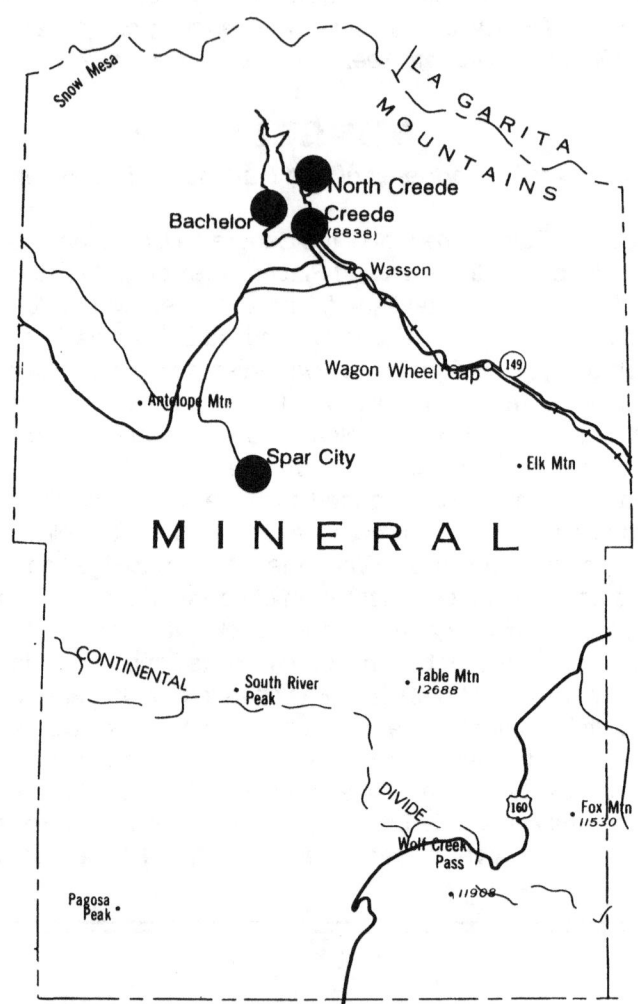

MINERAL COUNTY

"Be thankful for fools. Without them none of us would amount to much."

BACHELOR

Location: Near Creede, on the Bachelor Loop. The shortest route is to take the south loop (Road #504) west and then north from Creede for a distance of four miles.

In September of 1891, Mr. and Mrs. C. L. Calvin constructed a home and a boardinghouse, which were the first buildings erected in the town of Bachelor. Shortly thereafter, the Last Chance Mine struck paydirt, which coupled with other mining activity in the winter of 1891-92 created a bustling and rowdy boom town. Fortune seekers flocked to Bachelor Mountain by the hundreds, digging everywhere including the site of the planned city of Bachelor. After the chaos subsided another happening contributed to the population boom. Many citizens from nearby Creede moved to the high mountain town site after the disastrous fire in Creede on June 5, 1892. There seems to be no accurate source for Bachelor's peak population. Estimates range from 1,200 to as high as 6,000 during the boom period.

Following the Calvin's boardinghouse, the first three "businesses" to spring up were two saloons and a parlor house — a reflection of the wild nature of the community. Shootings, accidents, and fires were common occurrences. An excellent fire fighting brigade checked and controlled the fires, however, so the town avoided the type of devastation suffered by Creede.

A sad story is told about the town's reforming minister. One winter when his small daughter was dangerously ill with pneumonia, the minister traveled down to Creede to fetch a doctor. Upon returning home, he found a strange man hovering over his daughter's bed. Thinking the man was up to no good, he killed him on the spot. After learning the man was a doctor attempting to help his daughter, the minister was distraught and committed suicide. His daughter also died of her illness. A gravesite located below Bachelor in a grove of aspen is said to contain the bodies of the minister, his daughter and the doctor. Supposedly they are buried one on top of the other because of the difficulty of digging graves in the frozen ground of winter.

Like other mining communities there were parlor houses, gambling halls, and saloons. There must have been a little culture in Bachelor, however, for it had its own opera house, and the Bachelor City Dramatic Club which was touted as being excellent. The town also had its churches. The Congregational Church boasted that it was the highest church in the United States (the elevation of the town was 10,526).

The mines in the Bachelor area were among the richest in the Creede district. The boom and the glory were short lived, however. The silver devaluation in 1893 dealt Bachelor a severe blow. Although the town held on for several years, most of the residents moved off the mountain, down to Creede or elsewhere. Today, one partially roofed cabin, surrounded by trees, and the remains of two other cabins are all that exist at the site of Bachelor.

A letter sent back east said, "There are mighty few women here and they don't come close to handsome — but that's not held against them."

NORTH CREEDE
Location: 1 mile north of Creede

The area "between the cliffs" was the first hub of activity in the vicinity. It was originally known as Creede, then Upper Creede when Jimtown to the south became Creede. Eventually it was renamed North Creede.

The community, set in a narrow and deep gorge, was and is a dangerous place to live. Heavy rains and melting snow cause the water level to run high. Each year threatening torrents of water roar through the area. A flood in 1918 washed away the Cliff Hotel and the Holy Moses Saloon as well as the section known as Stringtown. North Creede was devastated. Nevertheless, several buildings which hug the cliff wall have resisted floods, fire and the test of time.

During its boom years, North Creede had a population of several hundred. Today there is only a handful of residents in this picturesque suburb of Creede.

Here's a land where all are equal —
  Of high and lowly birth —
A land where men make millions,
  Dug from the dreary earth.
Here meek and mild-eyed burros
  On mineral mountains feed.
It's day all day in the daytime
  And there is no night in Creede.

The cliffs are solid silver
  With wondrous wealth untold,
And the beds of the running rivers
  Are lined with the purest gold.
While the world is filled with sorrow,
  And hearts must break and bleed —
It's all day all day in the daytime,
  And there is no night in Creede.

—Cy Warman
Editor, Creede Candle

CREEDE

Location: 55 miles south of Gunnison on State Highway

Although a little prior prospecting was done in the vicinity, nothing spectacular happened until 1889. Nicholas C. Creede and his partner George L. Smith discovered float near the junction of West Willow and East Willow Creeks. They traced the source to the head of West Willow Creek, made a claim, and named it the Holy Moses. The men left prior to the ensuing cold winter, and returned in the spring of 1890. David H. Moffat, president of the Denver and Rio Grande Railroad, and a group of others, purchased the claim almost immediately, for $70,000. This triggered one of the great booms in Colorado history.

In 1891 Theodore Renniger, grubstaked by his former boss Ralph Granger, discovered what the assayer called the richest stuff he had ever seen. They staked their claim which they called the Last Chance. Nearby, Nicholas Creede located the Amethyst Mine. These two mines were the richest silver producing mines in the Creede district during the 1890s.

Several camps cropped up between the cliffs, up the canyon, and spread to the small valley south of the cliffs. Originally, the area of the canyon by the junction of West Willow Creek and East Willow Creek was considered Creede. The area south of the fork was called Stringtown because it was so narrow between the cliffs. As the population spread south beyond the cliffs, Jimtown was established as the commercial center of the area. Jimtown was the approximate location of Creede today. The area north of the forks, originally known as Creede, became Upper Creede, and later officially North Creede. Additionally, Weaver blossomed beyond North Creede on West Willow Creek; Bachelor was established high on the mountain above Jimtown; and Sunnyside sprang into existence over the hill to the west of Bachelor.

Speculators, miners, gamblers, and parlor girls came on every train to the over-crowded vicinity. The population swelled to about 10,000. If ever there was a red-hot town, Creede was it. It was a melting pot of strange, different, and interesting people. Soapy Smith, a slick and intelligent con-artist, and his gang ruled the city for a while. Smith opened the Orleans Club and declared himself boss of the underworld. Bob Ford (the name who killed Jesse James) built a dance hall and saloon called the Exchange. On June 8, 1892, Ford was slain by a shotgun blast from Ed O'Kelley. O'Kelley was convicted, imprisoned and then pardoned a few years later. Ford's body was buried on Boot Hill, and remained there for a few years before his family transported it back to Missouri. Shortly after the body was removed, another killer was buried in the same grave. Another saloon, which was owned by a Denver firm, was managed by Bat Masterson.

The parlor houses were operated by Lulu Slain, her friend the Mormon Queen, Slanting Annie, Lillie Lovell, and Rose Vastine, known as "Timberline" for obvious reasons — she was six feet two inches tall. These "beauties" prompted the *Creede Candle* of April 29, 1892, to write: "Creede is unfortunate in getting more of the flotsam of the state than usually falls to the lot of mining camps — some of her citizens would take sweepstake prizes at a hog show."

At one time or another Creede also had three cigar-smoking women gamblers — Poker Alice, Calamity Jane, and Killarney Kate.

The history of Creede follows the boom and bust pattern of many mining communities. In 1892 a fire started in a saloon, spread down Creede Avenue, and eventually destroyed most of the business district. The fire and the silver crash of 1893 created a mass exodus and spelled the end of the boom.

Creede today is a friendly town — proud of its rich mining heritage, and ready to share it with the many tourists who visit each year.

When the Free Coinage Hotel was converted into a community hall, second floor bedroom doors were removed which read: "Rose-$1.00; Marie-$1.50; Ruth-$3.00," etc.

SPAR CITY

Location: 14 miles southwest of Creede via State Highway 149 and Lime Creek Road

The picturesque camp was established in the spring of 1892 as Fisher City, after an early discoverer of float gold. It was also known as Lime Creek before being renamed for the Big Spar Mine.

By the summer of 1892 there were about 300 residents in the community, with many more in the surrounding hills. A saw mill was constructed. Main Street sprang up with the usual saloons, restaurants, and stores. *The Creede Candle* opened a newspaper office to publish *The Spar City Spark*.

A stage line was established between Spar City and Creede as well. The road to Creede ran over Robber's Hill, so named because it was a favorite ambush point for highwaymen.

A claim to fame for Spar City was a big pine bar which graced the principal saloon. Bob Ford, the man who gunned down Jesse James, was shot to death in his Exchange Club at Creede behind his own bar. A saloon keeper in Spar City, wanting the memento, purchased the bar and had it shipped by wagon with great difficulty to his establishment.

Spar City today consists of well-kept cabins, many of which still form a neat row down each side of Main Street. The community is private, fenced, and has a "No Trespassing — Members Only" sign at the gate. Special permission should be obtained before entering the premises.

MINERAL COUNTY 173

Townspeople line up for a photograph at Bachelor. *Colorado Historical Society.*

All the comforts of home and "no danger of fire" at this Creede hotel. *Colorado Historical Society.*

The Quinn brothers of Creede believed in signs. *Colorado Historical Society.*

Creede Avenue was a busy place. *Colorado Historical Society.*

NORTHWEST REGION

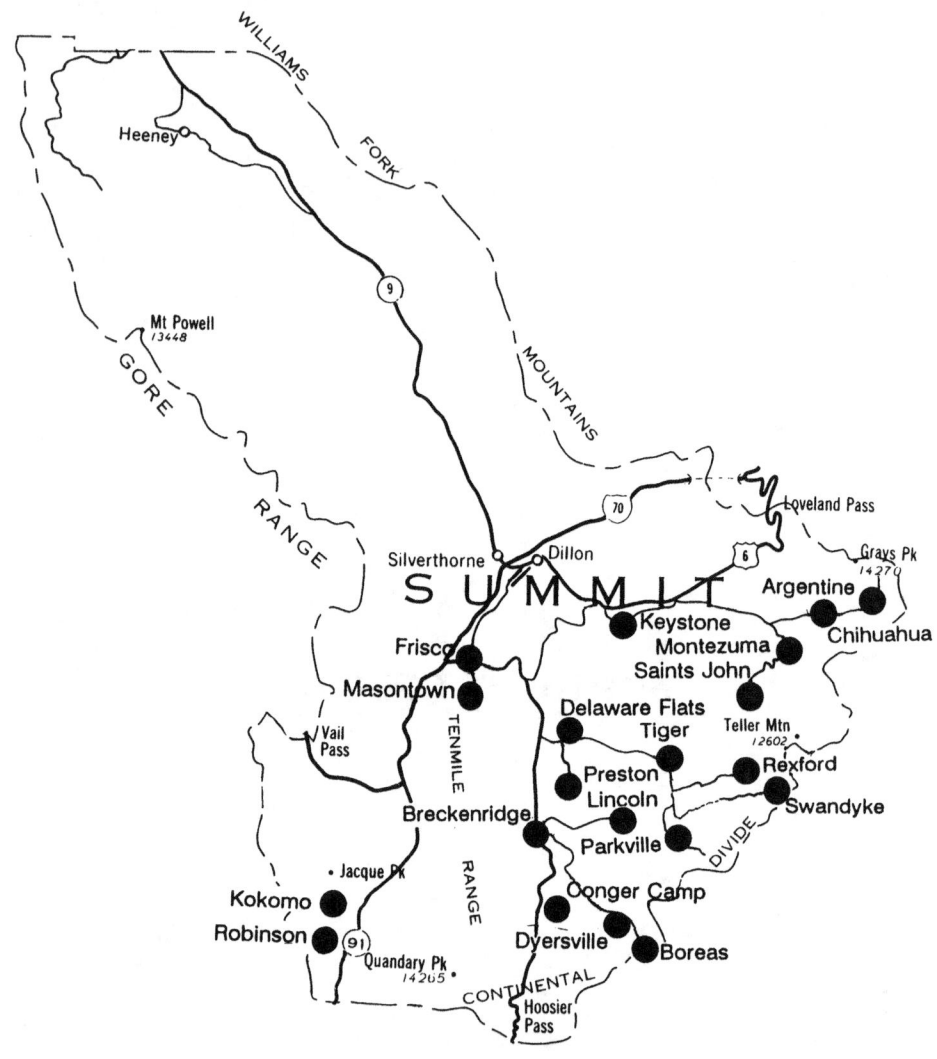

SUMMIT COUNTY

On one occasion, during the early days of Breckenridge, Judge Silverthorne ruled that two miners should settle their dispute by a duel. The two were to stand back to back, walk fifteen paces, then turn and fire. Instead of turning, they both ran in opposite directions, never to be seen again.

BRECKENRIDGE
Location: 10 miles south of Frisco on State Highway 9

Although there were a number of exploratory expeditions through the valley along Blue River during the 1840s and 1850s there was no permanent settlement until 1859. In August of that year, a group of southerners from Georgia and Alabama led by ex-General George E. Spencer descended into the area and began to pan the stream. History records that Ruben J. Spalding worked his first pan for gold that was valued at thirteen cents. His second pan was worth twice that. Another prospector in the group, William H. Iliff, washed out $7,000 worth of gold from a 40 square foot area across the stream. Convinced that they had just touched the surface, the group constructed a log fort for protection from the Indians, and the settlement that would become Breckenridge was started.

Spencer was instrumental in naming the new town after a fellow southerner, John C. Breckinridge, the current (Vice-President) of the United States during the administration of James Buchanan (President). This helped hasten the establishment of a post office, which occurred January 18, 1860.

During the ensuing year Buchanan softened in his sentiments toward the southern cause for slavery and lost favor with the north. In the Presidential Election of 1860 the new Republican Party candidate, Abraham Lincoln, won election over three other candidates — Democrat Stephen A. Douglas, Southern Democrat John C. Breckinridge, and John Bell of the Constitutional Union Party. Chagrin over the town's name had set in.

The southern sympathizers who founded the settlement were soon a small minority as tents and log cabins blossomed up and down Main Street. As the population swelled, eastern "Yankees", a large contingency in the community, subtly changed the town's name to avoid further embarrassment. The first "i" was replaced by an "e" and the name Breckenridge was born.

Stores, hotels and saloons sprang up. Breckenridge became the center of commerce as supply wagons rolled in and out of town. Population in the district was estimated at 9,000.

As nearby Parkville began to decline, the residents of Breckenridge wanted the county seat moved to their city. One night in 1862 a mysterious raid was made on Parkville and the county records all turned up missing. After the indignation resulting from this unjust act had subsided, the records were miraculously found and situated in the new county seat of Breckenridge.

As the easy gold from the placer mines declined, so did the population. Many of the young men left to join the Civil War, others left to prospect in more lucrative places. By 1866 the population of Breckenridge had dwindled to less than 500.

The early 1880s saw a revitalization in the community. With the discovery of gold in fissures and veins the mountainsides and gulches surrounding Breckenridge became sprinkled with new mines. Prosperity was further aided by a brief silver boom and abetted by securing a Denver, South Park and Pacific Railroad depot site in 1882. Log cabins were replaced by false-fronted stores and substantial homes many of which were trimmed with wood-lace or were of attractive Victorian architecture. By 1885 the population had risen to 2,000 within the town.

The first schoolhouse was constructed on Main Street in 1871. Almost from the outset this structure was insufficient. In 1882 a new two-story, four room school was built which served the community for over twenty-five years.

Almost every mining town during this era had its share of brothels, and Breckenridge was no exception. When you consider that for a long time the men in the community exceeded the women in population by a ratio of 30:1, the parlor houses and their "lewd" women offered a much needed outlet for the miners. When the men talked of "going over the blue," they were referring to the area west of the Blue River where most of the brothels were located.

By one account there were eighteen saloons in 1880. They all did quite well as there were still an estimated 8,000 people in the district.

The earliest boardinghouses were of a fairly temporary nature. In 1863 Judge Marshall Silverthorne built the first hotel. Many followed. The Grand Central was the largest, the Hotel Arlington possibly the most elaborate, but the Denver Hotel gained the most notoriety because it was robbed in 1898 by the infamous Pug Ryan and his gang. After a chase and a gun battle, Pug Ryan escaped.

One of the colorful characters of Breckenridge was Captain Samuel Adams. He had an absurd idea that he could set sail from Breckenridge down the Blue River and eventually locate a new water passage to the Pacific Ocean. The expedition departed in June 1869. After destroying four boats and two rafts the crew hiked back to Breckenridge. Adams was branded — "a preposterous, twelve-gauge, hundred-proof, kiln-dried, officially notarized fool."

Disaster, or potential disaster, often prompts action. In 1880 a forest fire singed the edge of town. A few buildings were damaged and everybody experienced quite a scare. A fire house was immediately constructed. In 1896 a more devastating fire occurred which burned both sides of Main Street between Adams and Washington. This time a water works was immediately constructed.

The winter of early 1899 brought unusually heavy snow. No train was able to reach Breckenridge for 79 days. As supplies in the stores ran low the town's people had to resort to a heavy diet of beef. When the train finally arrived in April everyone turned out to greet it in wild celebration. In addition to much needed supplies the six engines pulling two coaches delivered fifty bags of overdue mail.

Despite the gold dredging boom that created enormous rock piles in the vicinity, the population of Breckenridge steadily declined and was listed in 1900 as 976. After the dredges shut down in World War II it looked very much like a ghost town.

Today, Breckenridge again glitters with the prosperity it once had — now as a world-class ski resort.

During an election for County Attorney, candidate Tom Miller advised a Parkville crowd that ...as a County Attorney in Kansas, he only lost one case, and it was on a technicality. When a rival asked "What was the technicality?" Miller's reply was, "The mob broke into jail and lynched him." Miller was elected!

PARKVILLE
Location: 10 miles east of Breckenridge, via Tiger Road and the South Fork

For a few short years, Parkville was quite a city. After gold was discovered in Georgia Gulch in 1859, the rush was on. Estimates claim that as many as 10,000 persons lived in Parkville during the early '60s. Parkville became a mining center, supply town, the county seat, and hub of social activities in the vicinity.

Stage lines arrived into the community from Breckenridge to the west, and over Georgia Pass from the southeast. There was a county courthouse, post office, several hotels and restaurants, saloons, stores of various specialties, and a newspaper. The playhouse attracted the theatrical company of Jack Langrishe, of Central City fame, and other performers. One of the earliest Masonic Lodges was built here.

There was even a mint. J. J. Conway and Company stamped gold pieces in denominations of $2.50, $5.00, and $10.00. Samples of their coins can be seen at the Smithsonian Institution, in Washington, D. C.

In 1862 all of the county records mysteriously disappeared during a moonlight raid on the courthouse. Later, after all the commotion had subsided, the records strangely appeared at the new county courthouse in Breckenridge. By this time, the short life of Parkville had already peaked and was in decline. No attempt was made to retrieve the records.

The post office closed in 1866. Hydraulic mining during the '80s aided the destruction of Parkville. A few mining ruins, foundations, and the cemetery are all that remain.

DELAWARE FLATS
Location: 14 miles north of Breckenridge via Tiger Road

The camp at Delaware Flats was established in 1860 adjacent to the Swan River and east of its junction with the Blue River. It was home for many of the area's earliest prospectors who panned the rivers for gold. Placer mining occured all around Breckenridge, and the Delaware Flats area was no exception.

Following the turn of the century the camp disappeared. The dredges of Ben Revett and others moved in to cut wide paths through the valley totally revamping its terrain.

Today, developers are doing a magnificent job of reclaiming the land. New homes are cropping up throughout the area. Located nearby is the remains of one lonely dredge to remind us of the unique method of filtering used throughout the vicinity.

The postmistress of Lincoln had been drying some wire gold in her oven. While her back was turned, the mail carrier slipped some into his shirt. Before he could get it out, he had been "branded" for his sin.

LINCOLN
Location: 4 miles east of Breckenridge in French Gulch

Wire gold was discovered during the early 1860s by Harry Farncomb. He purchased several acres of land in the area which he worked for a while in relative obscurity. One day he carried a sack of nearly pure gold to a bank in Denver. Because he had not staked his claim in adherence with conventional mining practice, a group from Denver attempted to "take" his property. The action precipitated what is known as the "Ten Years War." What began as an expensive legal battle, ended in a gun battle. After seven hours of fighting, three men were dead and many others wounded. Nothing was settled. A third party eventually purchased the land. The Denver bank went broke — Farncomb became rich — and people moved into the gulch to get wealthy and live happily ever after.

The town that blossomed below the "Wire Patch" was originally named Paige City. Soon thereafter it was changed to Lincoln City but was more commonly referred to as Lincoln. By the late '70s Lincoln had a population of 300 and was growing. Some say there were 1,500 inhabitants in Lincoln in the '80s; others say that number included neighboring camps such as Wapiti and the whole gulch.

The first school in Summit County was built at Lincoln in 1862. Also that year, Father Dyer established the Blue River Methodist Mission. There were several business establishments — including two hotels. The Lincoln City Smelting Works processed the area's ore. A sawmill was also located at Lincoln.

Nearby, Colorado's largest solid gold nugget was extraced from French Gulch in 1887 by Tom Groves and Harry Lytton. The nugget which weighs 8 1/2 troy pounds was dubbed "Tom's Baby." It is on display at the Museum of Natural History in Denver.

In later years, Lincoln was home to workers on the dredging operations in French Gulch and employees at the Wellington Mine. Lincoln still has a few houses — and a few inhabitants. The homes look very different today, however, as they have acquired shingles, additions, and other appurtenances.

PRESTON

Location: Near Breckenridge via Tiger Road and Gold Run Gulch

North of Gibson Hill, near the south end of Gold Run Gulch, stand the remains of Preston. This was the town's location, regardless of what some earlier publications may have said.

Much rich mining was done in the vicinity of Breckenridge, and when some strikes were made on Gibson Hill in 1875, the camp at Preston was established. Other than the mining properties and a sawmill, the only business establishments were a general merchandise store and a saloon. Because of its close proximity, Preston was reliant on Breckenridge as a supply center.

The richest discovery near Preston was the Jumbo Mine, which was located in 1884. Shortly thereafter, the Jumbo Mill was constructed. Originally, the Jumbo was operated by Felix Leavick (for whom the town of Leavick, in Horseshoe Gulch, was named).

Although some mining continued into the 1930s, and intermittently thereafter, the town of Preston was all but deserted by 1900.

For those who are unfamiliar with the area, the turn off to Gold Run Gulch is somewhat disguised by the golf course. From Breckenridge, take State Highway 9 north 3 1/2 miles to Tiger Road. Turn east, then take the first dirt road south (which is immediately adjacent to the east end of the golf course). Follow this road south through Gold Run Gulch, past the huge skeletal remains of the Jessie Mill, and on to Preston. The remains of several cabins (one of which may have a mattress inside), mark the location of Preston. Several mining ruins are nearby.

TIGER

Location: 8 miles northeast of Breckenridge via Tiger Road on the Swan River

The Tiger Lode was located in 1864 by Corydon Swan Smith, D. W. Willy, and George Reed. The Hamilton, St. Cloud, and other discoveries followed. The Ball Mill was constructed, as was the town of Tiger, to house the employees of the mining properties.

The Royal Tiger Mines Corporation was organized and it consolidated most of the mines in the area. For the most part, Tiger became a company town. Employees were well provided for by the company. The community had a general store, water works, electricity, steam heat, and a doctor.

In 1918, when production was at its peak, Tiger was hit by a flu epidemic which caused many fatalities.

Ore had to be shipped out by sled much of the year, but, although it was slow, it was steady and it continued until the Royal Tiger stopped production in 1939. A few years later, dredges plowed up and down the Swan River. The town was burned to the ground in 1973.

SAINTS JOHN

Location: 9 miles southeast of Keystone; 13 miles southeast of Dillon

Saints John lies nestled in a picturesque mountain valley high above the town of Montezuma. The site was first named Coleyville after John Coley who located the first ore here in 1863. Freemasons renamed it Saints John four years later presumably for their patron saints John the Baptist and John the Evangelist.

Strikes were sometimes made in unusual ways, and the one made near Saints John by Bob Epsey was no exception. Suffering from a hangover one day, Bob decided to sleep it off. He laid down to take a nap beneath a shade tree. Upon awakening, he was still woozy and needed to grab onto a rock to steady himself. The rock broke off and there it was — solid ore.

Before long, Saints John became a one company mining town. The Boston Silver Mining Association gave way to the Boston Silver Company in 1875 and operated for three years until the Boston Mining Company took over.

The collapsed remains of a long building can be seen in the center of Saints John. It was a two and a half story boarding house which also housed the company offices. Important visitors were entertained at the superintendent's house which was lavish and decorated with massive European furnishings.

Bostonian morality dictated that Saints John should not have a saloon — and it didn't. This didn't stop the miners, however, who regularly traveled down the mountain to visit the saloons, brothels, and poker dens in Montezuma.

Bostonian culture dictated that Saints John should have a library — so it did, with three hundred and fifty volumes.

John Coley built the first silver smelting furnace in Colorado. Its ruins can be seen on the hill high above the old boarding house location.

Poor access, hard winters, and less silver caused the decline of Saints John. The post office was closed in February of 1881.

MONTEZUMA

Location: 8 miles southeast of Keystone; 12 miles southeast of Dillon

D. C. Collier and H. M. Teller made important strikes which led to the founding in 1865 of the silver-rich mining camp of Montezuma. The town was named after the famous Aztec chief. Montezuma was important to the many mines in the vicinity, such as the Silver King, the Tiger, and the Queen of the West. Difficult access led to slow growth in the area. The completion of the Argentine and Webster passes accelerated progress somewhat. By 1880 there were 800 residents. The community had a schoolhouse, post office, three hotels, a bank, a smelter, a saw mill, a weekly newspaper, and several brothels and saloons.

The one room schoolhouse, which still sits prominently on a slope just east of Montezuma's Main Street, operated from 1884 until 1958. It was not Montezuma's first schoolhouse, however. In 1876, midway between Montezuma and Saints John, a small school was built which was called the Halfway Schoolhouse. It quickly outgrew its usefulness, and a second school was built in Montezuma in 1880. Pupils from Saints John had to travel down the mountain each day to attend school in Montezuma. This schoolhouse also proved inadequate, and the larger schoolhouse was constructed.

Montezuma and the surrounding camps were very close socially. They joined each other for community dances, sports, and celebrations.

Father Dyer, the "snowshoe itinerant," prospected when he wasn't preaching. On one of his many trips to the silver camp, Father Dyer located a mine on Collier Mountain.

The devaluation of silver in 1893 was a shock to Montezuma, as it was to most of the silver camps. Many people left, but some stayed hoping for better days. Montezuma never totally became a ghost town and still has a few residents today.

ARGENTINE

Location: 12 miles east of Keystone; 16 miles east of Dillon

Across the western frontier, there were men who had come west to escape a past they wished to conceal. Most mining camps had some. Stephen Decatur Bross, once a professor in Poughkeepsie, New York, "disappeared" in the 1840s, leaving his wife and two children behind. Silver was discovered along Peru Creek, and a town was laid out in 1868 by Stephen Decatur (the Bross was gone). The town, which was named Decatur after its founder, later became Rathbone and finally Argentine. Prior to statehood, Decatur served the Territorial Legislature. His brother also had a propensity for politics and was elected Governor of Illinois.

When word reached Governor Bross that a man who called himself Stephen Decatur bore a strong physical resemblance to him, Bross was sure he had found his long lost brother. The governor traveled to Colorado and identified his missing brother. Decatur publicly denied any relationship. Regardless of the coincidence, the truth in the matter is still speculative.

The community phased in and out over the years. Federal postal regulations required that a post office once closed must open under a new name if reestablished — hence the name changes.

The best production — of several good producers — came from the Pennsylvania Mine discovered in 1879 by J. M. Hall. The mine's yield exceeded three million dollars.

Just north of town near the Peruvian Mine lay several cabins. The area was referred to as Peru. Peru was the site of a clever swindle. George A. "Gassy" Thompson and his workers were hired by some absentee mine owners to dig a 100 foot tunnel into a mountainside. Instead of digging into the mountain, Gassy began building snowsheds which "tunneled" out from the mountain. When the snowsheds measured one hundred feet, and were covered by heavy snow banked against the mountain, Gassy announced that his work was completed. After inspecting the project, the owners complimented Gassy on his work and paid him in full. By the time the snow melted, Gassy and his men were long gone.

Argentine never was a very large town. In addition to the post office, there were several stores, and a hotel — the Sautell. Children from Argentine attended a "community" school which was located midway between Argentine and Chihuahua, its neighbor.

Somewhere near the schoolhouse west of town lived a miserly old couple, the Mitchells. They worked a small claim but also had odd jobs to help put bread on the table. They hired and boarded a miner to help them work their claim. When payday rolled around, the Mitchells announced that his wages were equal to the boarding bill he was presented. The miner didn't stay another day.

Most everyone left after the silver panic of 1893. But the town rebounded some until 1898 when a massive avalanche flattened most of the buildings. The few residents that were left hung on for a while. The post office closed in 1907, and Argentine was abandoned.

"Hanging was — a suspended sentence."

CHIHUAHUA

Location: Near Montezuma. On the Argentine Road nearly 3 miles east of the Montezuma Road.

The life of Chihuahua was short lived. It was incorporated in 1880 and destroyed by a forest fire in 1889. During the interim, it boomed. There were two hotels — the Chihuahua and the Snively, a sawmill, a reduction works, several stores, and a small schoolhouse east of town which it shared with its neighbor Decatur (later to become Argentine).

Chihuahua had no doctor or preacher. The residents boasted that none were needed, so the story goes, because there wasn't any sickness or sin.

A story is told about two Chihuahua prospectors (the good guys) who were waylaid by several rogues (the bad guys). The prospectors were robbed and killed. Residents quickly heard of the tragedy, formed a posse, and went after the killers. Three of the rogues were caught and hanged on the spot. All five bodies were carried back to town. Somewhere near Chihuahua there are two gravesites — one for the good guys and one for the bad guys. The preacher who wasn't needed in this "sinless" town, wasn't there to give last rites.

KEYSTONE

Location: 5 miles southeast of Dillon via U.S. Highway 6

Keystone had no mines but was of immense importance to the ming industry in the mountains above it. The Denver, South Park & Pacific Railroad (later the Colorado & Southern) which left Como and climbed Boreas Pass eventually terminated at Keystone. Wagon roads from both Loveland Pass and Argentine Pass connected with the railroad terminus. Ore was shipped from many mining camps such as Montezuma, Saints John, and Argentine down to Keystone to be carried by the railroad to smelters on the eastern slope.

The rails were removed in 1937. Today Keystone is a popular ski resort. Some of the buildings from the "old town" are located on the property of the Keystone Science School. Of special interest are the cabins with exceptionally low eave heights.

MASONTOWN
Location: 1 mile south of Frisco

General N. B. Buford (Superintendent of the Federal Union Mine at Colona Bar, near Idaho Springs) staked a claim above Rainbow Lake in 1866, and other claims followed. Nothing much happened, however, until 1872 when a group of Pennsylvania investors constructed a reduction plant at a cost of $75,000. The settlement and the Masontown Mining and Milling Company were both named for their hometown in the east. Transportation from the site was difficult, and the mill was never very successful.

Masontown preceded its neighbor, Frisco, by several years. By the time Frisco began to grow, Masontown had declined.

Throughout its history the settlement was beset by avalanches. A slide in 1912 destroyed the mill, and another in 1926 wiped out most of what remained.

The site which is remote and well-hidden, experienced a revival during prohibition. Bootleggers constructed whiskey-producing stills which may have been more profitable than the minerals extracted during earlier years.

FRISCO
Location: 10 miles north of Breckenridge on State Highway 9

Frisco, today, is an attractive blend of much new and some old. The community and the Frisco Historical Society have done a fine job of preserving many old buildings while the town blossoms around them from the influx of tourists.

The town, which was originally a Ute Indian camp, was founded by Henry Recen in 1873. It is said that Henry Learned named the town when he tacked a sign above his cabin door which said "Frisco City," and the name stuck. Frisco was a mining town, although the immediate area was not heavily mined. When the railroads arrived the town seemed to step forward as a transportation center. Both the Denver & Rio Grande and the Denver, South Park & Pacific chugged into Frisco, which by 1884 had a population of about 250. Main Street had two hotels, many stores, several saloons, and was the center of activity.

Frisco never became a ghost town — but it tried. Population fluctuated with mining. After a couple of ups and downs there were only 18 residents by 1930.

Once a saloon in the 1890s, Frisco's original one-room schoolhouse stands on Main Street as a highlight of the Frisco Historic Park — an interesting complex of nineteenth-century buildings.

CONGER CAMP
Location: 3 miles south of Breckenridge via State Highway 9

Far below Dyersville, at the foot of Indiana Gulch is the site where Conger Camp once existed. The camp (sometimes called Conger's Camp or Conger) was named for Colonel Sam P. Conger (see Caribou) who located the rich Dianthe Mine. There were other good strikes which included the Case, Highline, Newark City, and the Franklin. The predominant yield was silver and copper.

A sawmill was constructed, as were several business establishments and about thirty or forty houses. The settlement was served by the Spottswood and McClellan Stage Line.

Conger Camp lasted about three years. Its short life is attributed to several factors. There were plenty of prospectors but the supply of miners was low. Outside investment could not be enticed and working capital was low. There were high expectations and low realizations — the mines just didn't develop as anticipated. A little lumber activity continued after 1882, but Conger Camp had virtually died.

A baby girl was born at Boreas in 1882 and hailed as "the highest born lady in Colorado."

BOREAS
Location: 11 miles southeast of Breckenridge at the summit on Boreas Pass Road

At the summit of Boreas Pass, stands the location of Boreas. Once a stop for weary travelers pushing their wagons from Como to Breckenridge, or vice versa, Boreas took on added importance with the completion of the Denver, South Park and Pacific Railroad narrow gauge track across this route in 1884. A depot, section house, engine house, and other buildings were constructed. Adjacent to the depot a 600 foot long snow shed was built. It was later extended to a length of 957 feet.

Stories of trains marooned in snow banks were commonplace. During the record setting winter of 1899, Boreas was isolated for ninety days. As supplies began to run out, two men set out for Como on snowshoes. Their frozen bodies were found the next day. On another occasion, a man on snow shoes left for Como to fetch medical supplies for his sick wife. When the snow melted the following summer, his body was found.

Even in fair weather there were sometimes problems. A runaway train with thirteen cars of ore derailed and crashed in 1901. A brakeman was killed. In another instance the Phineas T. Barnum Circus train just couldn't make it to the top of the grade. As a result of somebody's brilliance, the elephants were unloaded to push the train to the summit. The circus went on.

The post office which was established in 1896 was discontinued in 1905. The two-story section house and shed that stand at the top of the pass have been reconstructed.

DYERSVILLE

Location: 11 miles southeast of Breckenridge via Boreas Pass Road

The Warrior's Mark Mine was discovered in the early 1880s by the snowshoe itinerant, Father John L. Dyer (also see Buckskin Joe). Dyer built a cabin — then a couple of others. He hired a few workers to help him with his mine.

Neither the camp, which is a stone's throw from the summit of Boreas Pass, or the mines ever amounted to much. A fellow named Thompson, previously a boarder at Dyer's home in Breckenridge, staked the Thompson Claim nearby. It also was a poor producer.

Throughout his life, it seems as though Father Dyer was always "broke." To save on the cost of testing, he made a Breckenridge assayer a partner in his mining property.

Dyer moved his wife to the camp in 1881, but neither stayed very long. The retired preacher was nearly seventy and didn't have the energy he once had. The high-mountain weather was tough on the couple and they moved back down to Breckenridge. For his interest in the Warrior's Mark, Dyer ended up with about $2,000.

The Dyer cabin, and a few other ruins, can be seen at the camp which is located at the top of Indiana Gulch.

ROBINSON

Location: 15 miles south of Frisco on State Highway 91

In the late 1870s miners swarmed over the mountains looking for the same rich carbonate ores that created the boom in nearby Leadville. Some crossed Fremont Pass to the north and discovered deposits of silver. A tent colony emerged and was named Carbonateville. As it did in most all mining settlements, every new discovery brought in more prospectors. Nearby, Robinson's Camp was founded.

Two prospectors, Charles Jones and Jack Sheddon were grubstaked by a Leadville merchant, George B. Robinson. Jones and Sheddon had located several fine claims. It is estimated that two thousand people flocked into the immediate area. Robinson's Camp and Carbonateville fused to become the town of Robinson, named after its principal benefactor.

George Robinson, realizing the potential of the bonanza that was occurring, moved quickly to buy out his partners. Backed by some New York financing, the Robinson Consolidated Mining Company was created. Robinson became immensely rich and extremely popular. In November of 1880, he was elected to the office of Lieutenant Governor of Colorado. Within a month he died tragically. A dispute had emanated between Robinson and Captain J. W. Jacque regarding the ownership of the Smuggler Mine. Expecting violence, Robinson posted armed guards at the site. On the night of November 27, 1880 Robinson went to the mine to check on the guards. Thinking he was an intruder, one of the guards shot Robinson. He died two days later. According to the guard, he had called out to Robinson to identify himself but received no answer.

The town of Robinson, although it couldn't compete in size with rival boom-town Kokomo just a mile and a half to the north, did in fact become the principal business center for the county. Before his death, George Robinson financed the construction of a hotel, bank, and smelter. Another hotel followed, as did a Catholic Church. The first train arrived on New Year's Day 1881 amidst a glorious celebration. Later that same year a school was established.

Within a year progress had ended and decline had set in. The once rich mines were becoming exhausted. A fire in 1882 destroyed many of the town's buildings. The repeal of the Sherman Silver Purchase Act took its toll also. A few families stayed, but most left. The prosperity was over.

Robinson is now buried forever at the bottom of a tailing pond of the Climax Molybdenum Company. A marker along State Route 91 identifies the location.

So the story goes — an early discovery, the Dead Man claim, was made while one miner was digging a grave for another just deceased.

KOKOMO

Location: 14 miles south of Frisco on State Highway 91

Beside the highway on State Route 91, north of Fremont Pass, stands a monument which reads: "In this valley the towns of Robinson, Kokomo and Recen existed. Kokomo was the site of the highest Masonic Lodge in the U.S.A — Elevation 10,618 feet." This memorial and the tailing ponds of the Climax Molybdenum Company beyond are all that denote the location of a once active mining town.

Although there was some placer mining in the early 1860s, it wasn't until the boom in Leadville in the late 1870s that the Ten-Mile Mining District became industrious. Some very rich strikes were made, and each brought more people into the area. Kokomo, named for the city in Indiana from which many early residents came, blossomed into the largest town in the district.

The Kokomo post office was established on May 5, 1879. Less than a month later, the town was incorporated and became, at the time, the highest incorporated town in Colorado. Later the same year, the first issue of Summit County's first newspaper came off the press. *The Times* was distributed to Kokomo residents on September 27, 1879.

Adjacent to Kokomo was the town of Recen which was incorporated in 1880. It was named for three brothers, Andrew, Henry, and Daniel Recen who were important in the town's development. The railroad arrived at Recen in 1881 and Kokomo in 1882. The two communities gradually grew together and jointly were called Kokomo.

Much of the town was destroyed by fire in October of 1881 and the population gradually decreased after that. The panic which swept through mining communities in 1893 affected Kokomo as well.

After robbing the Denver Hotel in Breckenridge during the summer of 1898, the notorious Pug Ryan and his gang fled to a cabin near Kokomo. They were tracked down. In the bloody gunfight that followed, two lawmen and two members of Ryan's gang were killed. Ryan escaped, however. He was captured in Seattle four years later, escaped, was recaptured and died in prison in 1931.

A group of school children found part of the loot ten years after the robbery. While on a picnic they discovered it stashed in a hollow log near the cabin that was used as a hideout.

Kokomo never entirely became a ghost town until the remaining buildings were destroyed by the Climax operations in 1971.

"The mule's tail was the prospector's compass. All he had to do was follow it."

REXFORD

Location: 14 miles northeast of Breckenridge, via Tiger Road and the North Fork (4WD)

Daniel Patrick discovered the Rochester lode in 1880. The Rochester King and the Rochester Queen were established. In 1881, the Rexford Mining Corporation was organized to operate the properties. It was capitalized for $100,000.

Southwest of the Rochester, Rexford was built as a company town. The community had several business establishments which included a hotel, boarding house, saloon, general merchandise store, and even a gin mill. The mail carrier delivered from Montezuma to Rexford twice a week via the trail that skirted Glacier Mountain.

Rexford had a short life. The mining properties, which yielded about $5,000 per month during the early years, gradually dwindled.

The last structure, a false-fronted hotel, collapsed a few years ago. Amidst the picturesque meadow which is the site of Rexford are several foundations, partial walls, and piles of rubble.

SWANDYKE

Location: 14 miles east of Breckenridge via Tiger Road and the Middle Fork

Swandyke is located at a high elevation just west of the Continental Divide. It was a late-bloomer, which prospered in the 1890s. The gold camp was actually divided into two sections which are about a mile apart. The section located closer to the headwaters of the Middle Swan River is sometimes called Upper Swandyke. Stagecoach service connected Swandyke with Breckenridge, and across the Continental Divide with Jefferson.

The population of Swandyke peaked at about 500 during the mid '90s. A post office was finally established in 1898 (and discontinued in 1910, long after everyone had gone). The Summit House Hotel was the center of activity. There was also a boarding-house, general merchandise store and a few saloons.

Some of the mines in the Swandyke area were the Potter, Gibbs, Three Kings, Tyler, and Uncle Sam. There was a mill to handle the ores from the various mines. Most of the properties were operated by the Swandyke Gold Mining Company. The high altitude, remote location, and heavy snows made it expensive to ship ores out for further reduction.

Judging from the dates of old newspapers found in one of the cabins years later, 1901 was the last year Swandyke was inhabited. Buildings still stand at both locations. Several other foundations can be seen. At Upper Swandyke, below the head waters of the Middle Swan, lies the remains of a water wheel.

Those traveling to Swandyke (from Breckenridge) via the middle fork may be interested in the site of Middle Swan. When you reach the fork where the road to the right crosses the creek, check out the "cushioned" meadow just beyond which was the location of the old saw mill. The left fork continues on to Swandyke.

Looking west down Lincoln Avenue in Breckenridge about 1881. *Colorado Historical Society.*

There were several mining dredges in the vicinity of Breckenridge. *Colorado Historical Society.*

Jessie mining property in Gold Run Gulch.
Denver Public Library, Western History Department.

A five dollar gold piece die from the stamp machines at the mint of J.J. Conway and Company in Parkville. *Colorado Historical Society.*

194 NORTHWEST REGION

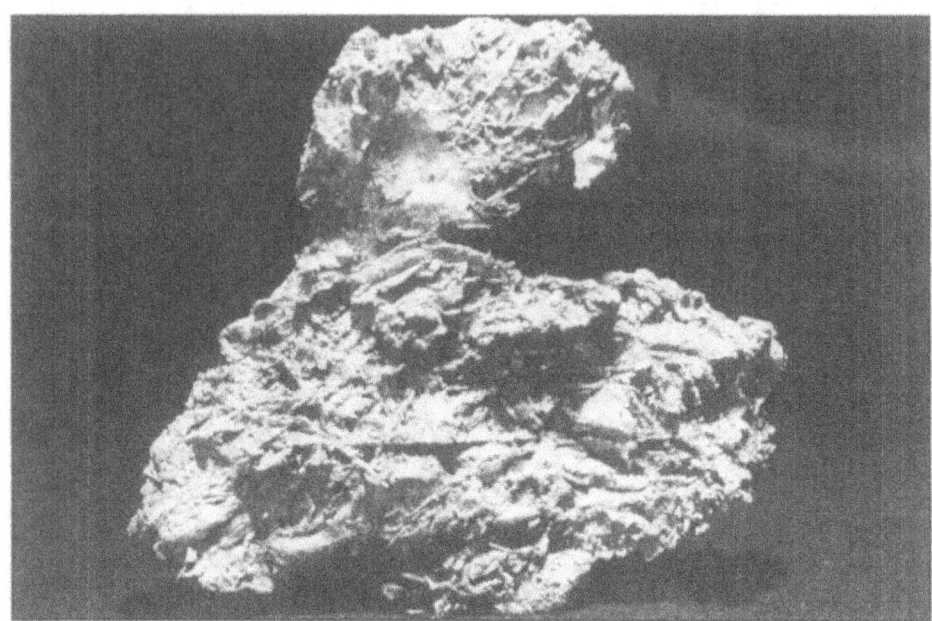

"Tom's Baby," Colorado's largest gold nugget (8 1/2 troy pounds) was extracted from French Gulch in 1887 by Tom Groves and Harry Lytton. *Denver Museum of Natural History.*

The community of Tiger disappeared into ashes. *Colorado Historical Society.*

The smelter at Saints John. *Denver Public Library, Western History Department.*

The early days of Montezuma. The Rocky Mountain House is at far left. The Summit House is at right center. *Colorado Historical Society.*

The high mountain community of Argentine. The store of Dick Hall is at left. Hall (wearing overalls) watches Frank Carman with the horses. *Colorado Historical Society.*

The Denver & Rio Grande pulls into Robinson. *Colorado Historical Society.*

SUMMIT COUNTY 197

Ten Mile Avenue in Kokomo. A trademark of the community was the modified false fronts which minimized snow-banking creating less live-load on the roofs. *Colorado Historical Society.*

Looking down Ten Mile Avenue in the opposite direction, the buildings of Recen can be seen in the background. *Colorado Historical Society.*

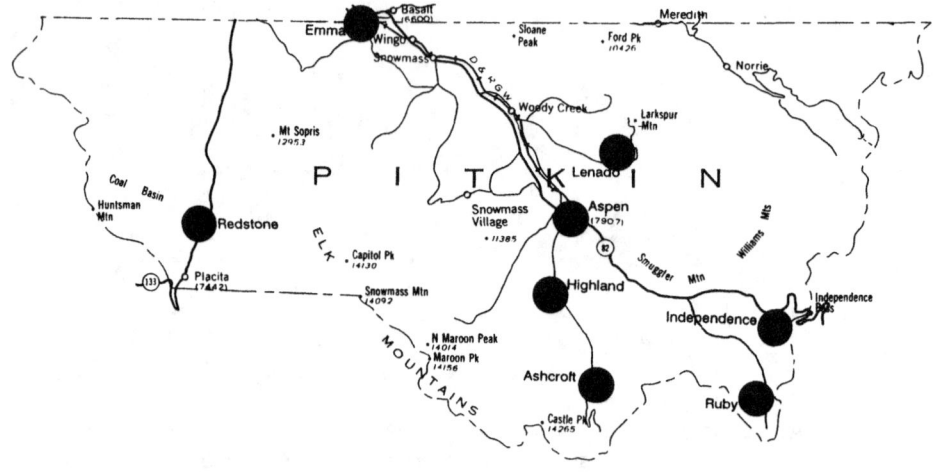

PITKIN COUNTY

INDEPENDENCE

Location: 17 miles southeast of Aspen on State Highway 82

The site where Independence once thrived lies in a beautiful meadow in the high mountain country just below Independence Pass. About five miles west of the summit the Independence Lode was discovered on July 4, 1879, and named for the day on which the find was made.

History records boundless confusion with regard to this community. Conflicting sources credit the discovery of the lode to different prospectors. Independence seems to be a merger of three or four mining camps which grew up around the Farwell Mine and the Independence Mine. Chipeta (named in honor of Chief Ouray's wife), and Sparkhill were separate camps and both had post offices. No post office was ever established here under the name Independence. During the boom years of the community, in the 1880s, business directories referred to both Independence and Sparkhill in order to avoid confusion. The settlement also changed its name to Mammoth City, and again to Mount Hope, before reverting to Independence once again.

The population peaked in the early 1880s at about two thousand people. Of the forty businesses, ten were saloons.

No railroad ever reached this mountain meadow. Its arrival into the valley below, however, aided the demise of the community, for Independence Pass became much less useful. As the mines played out, businesses packed up and moved and the stagecoach lines terminated their service to the area. There were only about one hundred people left by 1888.

Currently, the Aspen Historical Society, with limited funds, is doing everything possible to impede the deterioration of Independence.

When our grub pile's slim and scanty
Not a dollar in the shan
And our threadbare garment's letting in daylight;
The pay-streak sill eluding,
And barren dykes intruding,
And we are chased by harsh erectors day and night.
When our efforts lose their footing,
Our pard's insults sure cutting,
And misfortune's cruel jeers and sneers are keen;
From our Ashcroft habitation
We behold bleak desolation
When sear Autumn's goals' transform to silver sheen.

- Jack Leahy

ASHCROFT

Location: 12 miles south of Aspen via State Highway 82 and Castle Creek Road

In 1974, the Aspen Historical Society leased the townsite of Ashcroft from the U. S. Forest Service, for the purpose of preserving a part of history. Some of the buildings are original. Others were moved to the site to replace those which had deteriorated away, or were torn down by the Forest Service because they were unsafe. Although there seem to be some inaccuracies in the tabular history located at the site, the total effort is a valiant one. Ashcroft is definitely a ghost town worth seeing.

Following a few fairly good silver strikes in 1879 prospectors were lured into the vicinity. Castle Forks, as Ashcroft was originally known, was platted during the summer of 1880 by C. B. Culver and his group of miners. Many tent-dwellers moved from nearby Highland up Castle Creek to the new site. A post office was established on August 28, 1880, and John Nelson was named postmaster.

Ashcroft had five hotels. Four of those, the St. Cloud, the Riverside, the Fifth Avenue, and the Farrell, provided room and board. The Hotel View, which still stands at the south end of town, was never actually a hotel — it was a brothel.

The community had a newspaper, school, jail, doctor, bowling alley, haberdasher, several stores, and many saloons. The town even had a suburb — a cluster of cabins slightly upcreek — known as Hunley's Addition. Just north of town, on Express Creek Road (the road to Taylor Pass), precariously stands an old livery stable which was once used in an episode of "Sergeant Preston of the Yukon," a 1950s television serial.

In 1881, a couple of years before his marriage to Baby Doe, Horace Tabor and his partner Joe W. Smith purchased the Tam O'Shanter silver mining property near Ashcroft. Because it was difficult bringing ore out from its elevation of over 13,500 feet, the Tam O'Shanter was not a profitable endeavor for Tabor. He also had an interest in the nearby Montezuma Mine.

Horace and Baby Doe built a house in Ashcroft which was for a while a getaway from the gossip of Leadville society (see Leadville). For a few years after their marriage in March 1883, Ashcroft was their "summer retreat."

When the Independence Pass was completed, nearby Aspen began to grow — and Ashcroft began to decline. When the Denver and Rio Grande Railroad arrived in Aspen in 1887, the mountain passes above Ashcroft became useless. The population, which was at least 1,000 in 1883 (and estimated by some to be much more), dwindled to 150 before the turn of the century.

One man lived alone in the town for many years. Jack Leahy, known as the hermit of Ashcroft (see his poem above), finally moved in 1935, four years before his death. His cabin still stands north of the parking lot and entrance to the ghost town.

Over a hundred years ago, tramway towers on Aspen Mountain were used to haul ore down — today, newer towers haul skiers up.

ASPEN

Location: 42 miles southeast of Glenwood Springs on State Highway 82

The Leadville silver boom was such a bonanza that prospectors spread out hoping to find similar success elsewhere. During the summer of 1879 claims were staked on Aspen Mountain, West Aspen Mountain and Smuggler Mountain. By the spring of 1880 the rush to the Roaring Fork Valley was on.

Henry B. Gillespie was one of the first settlers. He purchased a few of the earliest claims and established a camp which he called Ute City. Gillespie had big plans for Ute City, but as snow set in he left for the winter. B. Clark Wheeler was more impatient, however, and challenged the snow. He had Ute City surveyed. Wheeler renamed the settlement Aspen, and with the assistance of David Hyman and others they platted an addition to the original townsite. The community was a crude mix of log cabins, frame structures, and many, many tents. The "business district" consisted of a hotel, restaurant, assay office, a few stores and other businesses, and several saloons.

The first road to Aspen came from Buena Vista across Taylor Pass and through the town of Ashcroft. Aspen became more accessible when Independence Pass from Twin Lakes and Leadville opened in late 1881. The three factors which contributed most to the early success of the town were the rich silver mines, financier Jerome B. Wheeler (no relation to Clark), and the railroads.

Some of the best early mines were the Durant, Smuggler, Spar, Mollie Gibson, Aspen, and the Castle. H. P. Cowenhoven, a merchant who would grubstake most any prospector, acquired a share of the Aspen Mine in settlement of a $400 debt. The mine made a fortune for Cowenhoven and his son-in-law David R. C. Brown. A prospector once sold a half interest in the Smuggler Mine for a burro and $50. The Smuggler produced millions, including the largest silver nugget in the world. Ore from the Mollie Gibson assayed at 3,300 ounces of silver per ton. The Durant which was purchased by David Hyman ran into the Aspen claim and wound up in litigation. The result was a compromise which made all parties rich. The Aspen was once leased by J. D. Hooper who made a rich strike shortly before the expiration of his lease. During the final days he extracted $600,000 worth of ore. Henry Tourtelotte (for whom Tourtelotte Park is named) located the Castle and several other lodes. Other top producers were the Midnight, Newman, Lone Pine, Washington, Consolidated, Argentum-Juniata, Park Regent, Montezuma, Free Silver, Bush Wacker, Vallejo, and the Emma. One small chamber in the Emma netted $500,000. The Lone Pine consolidated with the Mollie Gibson to form the Compromise Mining Company. The largest body of ore in the world found to date was located in 1883 in the Compromise Mine. During the same year, a new rich deposit was discovered in the Spar Mine.

Jerome B. Wheeler, a large stockholder in Macy's department store (New York) and the Colorado Midland Railroad among other things, arrived in the Roaring Fork Valley in 1883. He invested much money into Aspen and encouraged others to do likewise. The first electric tramway ever constructed for mining purposes was built by Wheeler. It carried ores from the Aspen Mine, down Aspen Mountain into the valley. In 1884 Jerome Wheeler built a smelter at the mouth of Castle Creek, and ores could be treated locally. Wheeler and Henry Gillespie built a mill and concentrator, and expanded underground exploration at the Mollie Gibson Mine. His investments and contributions were not always mining-related. He established and edited a newspaper, the *Aspen Daily Times*, and backed the construction of both the Hotel Jerome and Wheeler Grand Opera House. Both structures were impressive, and cost $1,000,000 each. The three-story hotel was lavish. Its furnishings were imported from overseas. The hotel had electricity, a barber shop, billiard parlor, and an elaborate dining hall. The opening Grand Ball in 1889, was a celebration of most luxurious splendor. The Opera House boasted exquisite ornamental woodwork with brass trimmings as well as plush upholstery and curtains. Wheeler was also instrumental in bringing the Colorado Midland Railway to Aspen.

A spur of the Denver & Rio Grande narrow gauge reached Aspen from Glenwood Springs in October 1887. The Colorado Midland, a standard gauge, arrived the following year from Leadville — through the newly constructed Hagerman Tunnel (which cost $2,000,000). A confrontation occurred between the two railroads which nearly wound up in gunfire. Mayor Henry Webber stepped in, and a peaceful solution emerged. Each railroad built separate depots and did so on opposite sides of town. They both shipped ore successfully for many years.

The arrival of the railroads boosted mining, which in-turn increased population. Mining output in 1888 exceeded $7,000,000, approximately double what it was in 1884. By 1889 it had reached nearly $10,000,000. Likewise, Aspen's population was 3,500 in 1884. Following the coming of the railroad its population expanded to 8,000 in 1888. The increase continued, and by 1892 the population had reached 12,000 and Aspen was the third largest city in the state. By then it had surpassed Leadville to become the world's greatest silver mining camp.

The crude little settlement of the early 1880s blossomed into a well-established city. It grew as a refined city — with class. Much of its exceptional society can be attributed to the quality and character of Aspen's forefathers. From the early years on, there were many churches, cultural activities, socials, lodges and civic clubs which projected for Aspen a much better image than that of most mining towns.

From February 1881 when Aspen was designated as the temporary county seat of newly created Pitkin County, politics has had its place. Soon after construction began on the new Pitkin County Courthouse in July 1890 controversy arose amid newspaper reports of bond fraud and misappropriations by county commissioners. Regardless, it was completed in January 1891. It is one of the oldest courthouses in Colorado still used as such. Davis Hanson Waite settled in Aspen in 1881 and launched the *Aspen Union Era*, a weekly newspaper which championed reform and Populist ideas. He was elected Governor of Colorado on the Populist ticket in 1892. After the silver crash in 1893, Waite received the nickname "Bloody Bridles," and became embroiled in controversy following his speech, " ... for it is better, infinitely better, that blood should flow to the horses' bridles rather than our national liberties should be destroyed ... " (*Rocky Mountain News*, July 12, 1893).

In addition to Waite's weekly newspaper there were at least two other weeklies and three dailies — the *Aspen Times* (*Aspen Daily Times*) a fine newspaper which was once purchased by Waite, then sold to B. Clark Wheeler (who later became his son-in-law), the *Rocky Mountain Sun*, and the *Evening Chronicle*.

By 1892 Aspen was a bustling city with electric lights and telephone service. Ten passenger trains arrived and departed daily. There were horse-drawn street cars. The Hotel Jerome was the finest of many hotels and boardinghouses. Besides the Wheeler Opera House, another popular spot was the Rink Opera House. Aspen had ten churches, three banks, a hospital, three schools, and even a racetrack. There were government buildings such as Armory Hall (National Guard) and the aforementioned Pitkin County Courthouse. Additionally, the business district was full of stores and specialty shops (such as the Kobey Shoe and Clothing Company and Reide's City Bakery). As with most every mining town, there were dancehalls, gambling dens, and saloons, the majority of which lined Cooper Street. They were supplied by Aspen's own brewery. Part of Durant Street was a "red light district." Many brothels housed the ladies of the night (and day). The city taxed each girl $5 per month. On weekends, large crowds would gather to watch Aspen's semi-pro baseball team take on teams from Denver, Leadville, and other cities. Homes throughout Aspen displayed much diversity. There was a multitude of miner's cottages, but there were also many stately houses — elegant examples of Queen Anne architecture.

The silver crash in 1893 had a devastating affect on Aspen. Within one month after the devaluation all area mines had shut down. Economic strife hit the community. Some millionaires had diversified their interests enough to stay afloat — others went bankrupt. After the price of silver stabilized, some of the mines reopened. The economy wouldn't rebound however. Aspen's population dwindled — and continued to do so. By 1930 the community had 700 residents.

During World War II, ski troops of the 87th Mountain Infantry (which later became part of the 10th Mountain Division) trained on slopes near Aspen. Several of the members loved Aspen and returned after the war to establish skiing. Walter Paepcke (head of the Container Corporation of America) envisioned an opportunity, and poured investment capital into the community and the ski industry. Aspen today booms as it did over 100 years ago, and is a world class ski resort.

"A blind mule can see just as well from either end."

HIGHLAND

Location: Near Aspen, 5 miles south of State Highway 82 on Castle Creek Road.

When several silver strikes were made in the surrounding mountains in 1879, the Highland Town Company laid out a city about six miles below Ashcroft to the north. They successfully lured many people into the area. Even at its peak, with 300 residents in 1880, Highland remained predominantly a tent colony. The future didn't look bright enough to build too many permanent structures. Several mineral pockets kept things going for a couple of years, but because no good veins were discovered, the boom was over by 1881.

Some activity continued in the area after the turn of the century. The Highland Tunnel was bored for a distance of about three miles in 1910 by the Hope Mining, Milling and Leasing Co. The project was abandoned within a few years, however, because the ore brought out was well below expectations.

LENADO

Location: Near Aspen - via Woody Creek Road - approximately 10 miles east of either Woody Creek Canyon turnoff from State Highway 82

Lenado (pronounced with a long a) blossomed into existence in the 1880s when A. J. Varney discovered lead and zinc ore on Porphyry Mountain. Other claims followed.

The town was inhabited by about 300 people at its peak, with half of them employed by Varney. There was a boardinghouse, a mercantile store, two saloons, a post office and many cabins. Later a sawmill was erected.

In 1888 the Denver and Rio Grande graded a road bed with plans to construct a spur to Lenado. No tracks were ever laid.

Today about one half of the many cabins are still inhabited. Abandoned vehicles litter the area, however, outnumbering the cabins about 3 to 1.

"Some towns are so small that when the train pulls into the station it's already out of town."

EMMA

Location: 2 miles southwest of Basalt

This site should not be confused with an insignificant camp named Emma which was located near Texas Creek in Gunnison County.

Robert M. Morrison established a small store on the stage line to Aspen, and the town of Emma was born. Following the arrival of the Denver & Rio Grande Railroad, a railroad station, section house, and water tank were rapidly constructed.

A post office was established on February 15, 1888. Emma Davis Garrison was named postmistress. She often cooked for the railroad workers and it was because of her overwhelming generosity that the town was named in her honor.

Later, in 1898, a general store and hotel were constructed of brick by Charles Mather.

RUBY

Location: Between Aspen and Independence - 12 miles south of State Highway 82 on Lincoln Creek Road

The remote site of Ruby (and the Ruby Mine) is situated high in Lincoln Gulch just west of the Continental Divide. From the time mining first began in the 1890s, it was very difficult to pack the ore down through the mountains. Initially burros carried the ores across the divide down Peekaboo Gulch to the two stamp mills of C. M. Everett, which were located at the junction of the North Fork and South Fork of Lake Creek at the road from Independence Pass. The crushed ores were then carted by wagon to "Smelter Valley" south of Leadville. An assay office and mill were constructed adjacent to the Ruby Mine which "lightened" the load somewhat, but shipping was still an arduous ordeal.

After the turn of the century a road was completed to Ruby from the opposite direction. Where Lincoln Creek intersects the road to Independence Pass, there stood a stagecoach station called the Junction House. It became the nearest feasible access to civilization for residents of the isolated camp. From the site of the Junction House, the road climbs along Lincoln Creek up to Ruby. The road was almost too late, however, as the community had already began to fade.

Below the mine, cabins once extended for a half-mile down Lincoln Gulch. In addition to the aforementioned buildings, there was also a boardinghouse, general merchandise store, and a large stable. Today, most everything has vanished. The old stable still stands. Three new cabins have been constructed

— much in the mold of those they replaced.

The 1900 census takers skipped Ruby as they did several other remote camps. It is estimated the population peaked at about 300. The camp was briefly called South Independence before its name was changed to Ruby — named for the mine, and its ruby silver.

Although predominantly a silver producer, the Ruby Mine yielded several other minerals including a little gold. There has been some mining activity off and on during the twentieth century.

REDSTONE

Location: 17 miles south of Carbondale on State Highway 133

John Cleveland Osgood, Chairman of the Colorado Fuel and Iron Company, built Redstone as a company town. He created a utopian community. The tudor-style Redstone Inn was constructed as a club house for Osgood's workers and a residence for the unmarried men. Unique, pastel-colored homes were provided for workers with families. Culture and education were of prime importance.

Alongside the Crystal River on 450 acres of land, Osgood constructed his magnificent 42-room mansion "Cleveholm." Some of the ceilings were gold-leafed. Oak wall panels were stenciled by Italian artists. Elaborate furnishings from throughout the world filled the English tudor manor.

Osgood had three wives — each of which created her own share of gossip. His first wife wrote steamy novels about their marriage. The life of his second had been touched with scandal. Because of her generosity, however, she was dubbed "Lady Bountiful." Osgood's third wife — who eventually sold Cleveholm after his death — was fifty years younger than he.

Osgood's C F & I operated coal mines at nearby Coal Basin for many years. A narrow-gauge railroad transported the coal down a grade, that sometimes exceeded four percent, to the coke ovens at Redstone for carbonization. The ovens, which were constructed in the late 1890s can be seen today opposite Cleveholm on State Highway 133.

PITKIN COUNTY 207

The Theodore Blohm Merchandise Co. at Aspen in 1888.
Colorado Historical Society.

This W.H. Jackson photograph shows Aspen from Aspen Mountain.
Colorado Historical Society.

John Cleveland Osgood constructed the Redstone Inn as a clubhouse for his workers and a residence for the unmarried men. *Redstone Inn.*

Alongside the Crystal River at Redstone, Osgood built his magnificent 42-room mansion "Cleveholm". *Redstone Inn.*

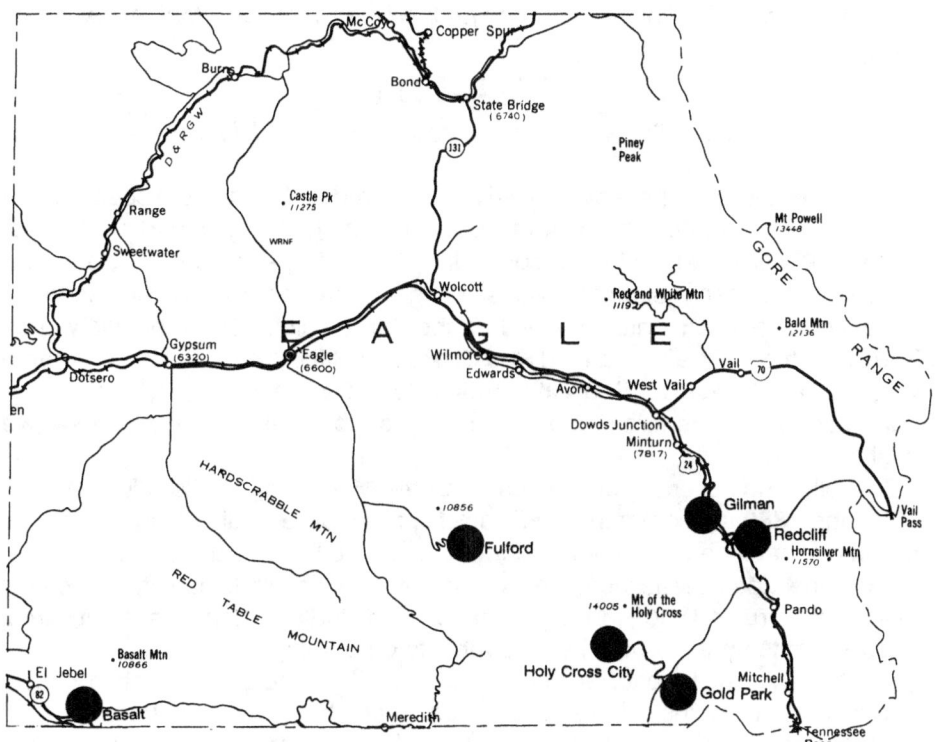

EAGLE COUNTY

A distressed and tearful dancehall girl known affectionately as "Big Hat" decided one day to "end it all." She ran to the bank of Turkey Creek and threw herself in. Her effort was futile, however, for she jumped into water only four inches deep.

REDCLIFF

Location: 22 miles north of Leadville via U.S. Highway 24

Redcliff is also spelled Red Cliff (as two words). In fact, both spellings are common. The sign on the post office contradicts highway department signs, maps, and books, which in turn contradict other maps and books. This author asked a half-dozen residents which spelling was proper. Three indicated that it should be spelled as one word — the other three said it should be two words. The post office officially changed Red Cliff to Redcliff in 1895. In 1979 the name was changed back to Red Cliff. Today, forestry maps, signs, and recent publications more generally use Redcliff. It probably matters little which way we spell it.

Following several good strikes on Battle Mountain and Horn Mountain in 1879 and 1880 the community of Redcliff sprang to life. It did so on both banks of Turkey Creek near its intersection with the Eagle River. During its early days, news of the Meeker Massacre reached Redcliff. Fearing an Indian attack, settlers quickly constructed a log fort. Apprehensions dwindled when no assault came, and residents again turned their thoughts to other things.

It wasn't long before Redcliff had three hotels — the Star, Mountain House, and Southern — plus a bank, post office, schoolhouse, a few stores, several saloons, and William B. Thom's controversial newspaper — the *Eagle Valley Shaft*. When Redcliff was snowbound during the winter of 1884, the remaining stock of paper was used up. Rather than stop the presses, Thom printed the news on wallpaper. The Star Hotel was so primitive that it had cloth partitions separating many rooms. Even so, the hotel imported a chef and was known to have the best "chow." Redcliff also had an opera house, a brass band, and a "restricted" cemetery. The first man buried there was killed by a bear. Another was crushed by a tree. However, when a few folks from nearby Astor City brought in the bodies of two men who had killed each other, they were refused admission on the grounds that murderers would desecrate their cemetery.

From its inception until recently, the economy of Redcliff has been reliant on Battle Mountain mines. Many minerals have been taken out, including silver, gold, copper, lead, and zinc. Some of the ore was processed through the Belden Smelter nearby. When the Denver & Rio Grande Railroad reached the area in 1881, much ore was shipped out to more efficient smelters.

The population of Redcliff peaked at about 400 during the mid-1880s. When Eagle County was established, Redcliff became county seat — a distinction it held until 1921. Since its beginning the community has always been inhabited.

With most mining discontinued and little else to fall back on, the economy in Redcliff has been down during recent years.

"Camp cooks are stove-in prospectors — too gimpy to work the mines."

GOLD PARK

Location: 28 miles northwest of Leadville on Homestake Creek

During the year 1880, prospectors found Homestake Creek to be rich in placer gold. An abundance of float gold was also found in the adjacent mountains. The rush was on.

Just one year later, Gold Park boasted two hotels, the Gold Park and the Homestake, a general store, post office, several saloons and a stamp mill. Population of the settlement had swelled to 400. A daily stagecoach connected with the Denver and Rio Grande Railroad at Red Cliff.

The Gold Park Mining Company, with its many claims, was the largest producer in the area. The company operated two mills, one west of Gold Park, the other above it at Holy Cross City. The two mills were connected by an iron flume two and a half miles in length.

Just two years later, Gold Park was nearly deserted. Production decreased to a point where additional investments of time and money were not warranted. Most everyone packed up and left.

The remaining buildings at Gold Park are on private property, fenced, with "No Trespassing" signs. They are located a short distance behind the gate which is opposite the parking area at the base of the Holy Cross City trail. Permission can and should be obtained before entering the premises.

"The higher you climb, the more rocks you have to dodge."

HOLY CROSS CITY

Location: 4 miles northwest of Gold Park (4WD)

It is said that gold was found in "twelve foot crevices with eighteen inch veins." Holy Cross City and its neighbors Gold Park, Missouri Camp, and Camp Fancy sprang into existence as the Holy Cross District at about the same time as the Leadville boom. Much money was invested into the area during 1881 and 1882.

The townsite lies in a picturesque meadow close to timberline. Its altitude of 11,335 feet meant harsh winters. Its accessibility was difficult. Both of these factors contributed to the short life of the town.

Holy Cross City had one hotel — the Timberline, the Holy Cross Mill, a school, an assay office, a general store which housed the post office, and many cabins. The population peaked at about 300 people.

By 1883 the quality and quantity of ore had deteriorated to the point where the mills closed down and the property was deserted. By the end of 1883, the only residents of Holy Cross City were H. W. Roby, manager of the Gold Park Mining Company, and his family.

A couple of dilapidated cabins mark the site of Holy Cross City. It is wise to wait for the high mountain snow to thaw and the run off to subside before making the climb up the steep grade from Gold Park to the location.

Though not visible from the townsite, three and one-half miles to the north (on a direct line) is the famed Mount of the Holy Cross, for which the town is named. A large snow cross lies in deep perpendicular crevices which preserve it throughout the year. The vertical column is over a quarter-mile in height, and the arms stretch over four hundred feet across (tip to tip).

BASALT
Location: 18 miles northwest of Aspen via State Highway 32

The trout fishing here is so great they used to say, "Out of the river and into the frying pan" — hence the name Frying Pan River, and the original name of the town Frying Pan City. The Colorado Midland Railway built a construction camp for railroad workers, a station, and railroad yard near the town and called it Aspen Junction, a name which stuck for several years. Several saloons, restaurants, stores, and a boardinghouse were constructed. It established a reputation of being a wild and rowdy railroad town with its share of gamblers and sexy ladies.

A tragic train wreck occurred at Aspen Junction on July 18, 1891 when a Midland locomotive collided with an excursion train. Hot steam from a broken check valve filled one car killing ten persons and badly scalding many more.

Because of nearby Aspen and ongoing confusion with its name, in 1895 the community adopted the name Basalt for the peak which rises above it. Basalt today is a pleasant small town and popular tourist stopover.

GILMAN
Location: 26 miles north of Leadville on U.S. Highway 24

Gilman is perched on a plateau 1,000 feet above the Eagle River. The setting is spectacular and can be viewed for some distance from each direction on U.S. Highway 24 (once Kelly's Toll Road). Today, the site is uninhabited, fenced, and is posted "no trespassing." It is the property of Paramount Pictures Division of the Gulf and Western Corporation, and according to rumor can be purchased for the "right price." The area has been designated an Environmental Super Fund Cleanup Site.

The mining camp began in 1879 and was called Rock Creek, Battle Mountain, and Clinton, before its name was changed to Gilman in honor of mine owner Henry M. Gilman. Activity actually started when the prosperous Belden Mine was located on May 5, 1879. The rich silver producer was named for its discoverer, Judge D.D. Belden. Shortly thereafter another lucrative silver find, the Iron Mask Mine was developed by Leadville newspaperman Joseph Burnell. Another silver property was the Eagle Mine. The Ground Hog Mine was a gold producer. Aerial trams carried ores from Battle Mountain to the large mill and railroad in the valley far below the townsite. Ores were refined, then carried away on the Denver & Rio Grande.

The population in 1899 was about 300. There was a weekly newspaper — the *Gilman Enterprise*. The community sponsored a drama club which provided cultural entertainment for all. The business district consisted of several boardinghouses and hotels, and a variety of stores. A fire which occurred that year (1899) destroyed the Iron Mask Hotel, the town's finest, as well as the schoolhouse, a mining building, and a few stores and houses. Mining continued through much of the twentieth century during which time Gilman was a company town and property of the Empire Zinc Company of New Jersey. Gilman may be viewed from the highway above town, but permission should be obtained from the Gulf and Western Corporation before entering the premises.

Nolan Creek, at Fulford, is named in honor of the prospector who accidentally shot himself to death while crossing a log over the creek.

FULFORD

Location: 22 miles southeast of Eagle via Forest Routes 400, 415, and 419

The lost fortune of Buck Rogers is perhaps Colorado's most authentic tale of buried treasure. In 1849 Buck Rogers and a group of prospectors left Illinois to join the gold rush in California. During their trek through present-day Colorado the party found traces of gold near a peak they called Slate Mountain. While the others ventured on, six of the prospectors, including Buck Rogers, remained to work the area. Their efforts were rewarded when a rich vein was located. Gold was extracted which they estimated as worth up to $100,000. The ore was stored in a drift until such time as it could be packed out. When provisions ran low, Rogers set out with $500 in nuggets and dust for the nearest camp 150 miles away. The others remained in what may have been Colorado's first gold camp. After struggling through blizzards for many days, Rogers reached his destination. Winter storms and warm saloons delayed his return trip for several weeks, but finally he departed. Upon returning to the location of the strike, Rogers was horrified at what he saw. An avalanche had carried snow and mud down the mountainside — burying the camp, men, and the gold as well. Following the incident, Rogers became a mental wreck — and he tried to drown his conscience in whiskey. He spent his remaining years wandering from one saloon to another telling his woeful tale to all who would listen.

Forty years later amid new discoveries in the area, a settlement cropped up which was later to become Fulford. The camp was so named following the tragic death of Arthur H. Fulford — which is another chapter in the Buck Rogers' story. While operating a stage stop on the road from Eagle, Fulford had a visitor who claimed to have found human bones, tools, and a treasure of nuggets at a location which sounded much like the description of Slate Mountain. He had covered his discovery and was looking for a partner to help him pack out the gold. Fulford agreed to help. As they prepared for their trip, a strange twist occurred in the story — the visitor was killed in a saloon brawl. For months Fulford combed the gulches of the East Fork of Brush Creek, but to no avail.l One day, while prospecting on nearby New York Mountain, fate dealt him the same cards that it did Buck Rogers' group years earlier. On New Year's Day 1892 an avalanche swept Fulford to his death. During the ensuing years, many prospectors combed the area up East Brush Creek. As far as we know, no one has ever found the location of Buck Rogers' strike — but, possibly one day someone will. Meanwhile, each time the story is told, the legend grows.

Gold, copper, silver, and lead were mined in the vicinity of Fulford. The Polar Star Mine and the Cave Mine were top producers. Other mines with a good yield included the Mendota, Kittie B., New York, Adelaide, Layton, and later the Lady Belle — a silver property. A twenty-five-stamp mill was constructed from which ores were transported to Eagle, site of the nearest railroad. A huge arastra — known as "The Wheel" — was built in 1893 or 1894 on Fools Peak Trail, not far from town. Access to some mines occurred through cave entrances. The Fulford Cave, one of many in the vicinity, remains a point of interest for tourists today.

The town of Fulford actually emerged as two separate communities — Upper Town and Lower Town. Lower Town was the larger, but both had hotels, boardinghouses, stores, saloons, and many cabins. The false-fronted Lamming Hotel — the largest, the schoolhouse and post office were in the lower section while the livery barn and assay office were above. Below New York Mountain on a small flat was a little settlement simply called New York Cabins.

Redcliff still looks much like it did in this early photograph. *Denver Public Library, Western History Department.*

This group is on the front porch of the Timberline Hotel at Holy Cross City. *Colorado Historical Society.*

Basalt's Midland Avenue was once called Railroad Street. *Colorado Historical Society.*

The Larson Hotel at Hahns Peak, about 1898. The business was owned and operated by Charles Larson. *Colorado Historical Society.*

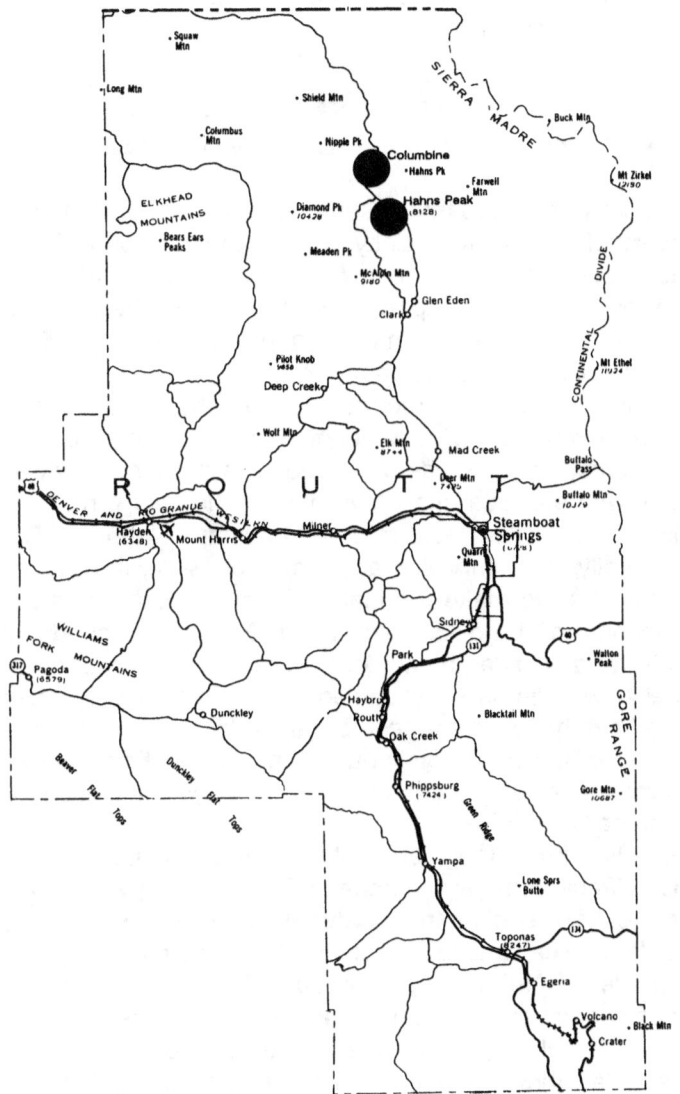

ROUTT COUNTY

The Declaration of Independence holds the truth that "...all men are created equal." Sam Colt assures it.

HAHNS PEAK

Location: 27 miles north of Steamboat Springs on State Highway 129

It is generally believed that Joseph Hahn, a German immigrant, made the first discovery of gold in the vicinity late in the year 1864. Hahn left for the winter, returning the following spring with his partners George Way and William Doyle. They established a company which was headed by Hahn, and organized a mining district as well. The town was platted, and by late 1865 several cabins were constructed, replacing the original tents.

With a few substantial structures erected, the three partners decided to remain in Hahns Peak during the winter of 1865. Before the snows came, George Way and the other town members headed to civilization. Way was to return to Hahns Peak with supplies and provisions for the cold winter. Heavy snows came, however, and Way never returned to camp. Hahn and Doyle were left stranded. Finally, after surviving as long as possible, the two men, sick and starving, attempted to make the long trek south. Hahn died during the trip. Doyle was found snow-blind and nearly dead. Despite this tragedy, the community became a thriving gold camp and, for a while, the county seat.

Actually, two camps were settled a short distance apart. One, known originally as Poverty Bar, later became Hahns Peak. The other, a short distance away, was called Bugtown or National City at different times. It was a dry camp which allowed no drinking or gambling. Bugtown was reliant upon Poverty Bar, the larger of the two camps.

Hahns Peak had a tough and determined sheriff named Charles Neiman. In 1898 Harry Tracy, formerly a member of Butch Cassidy's gang, and his partner David Lant were arrested for the murder of a deputy, Valentine Hoy. They were brought to Hahns Peak and locked up in jail to await trial. Tracy and Lant tricked and overpowered Neiman and escaped leaving the badly beaten and unconscious sheriff behind. Neiman regained consciousness and set out after the escapees. Realizing the fugitives were not dressed for the weather, the sheriff was certain they would try to hold up the Hot Sulfur Springs stagecoach. Nieman and his deputy boarded the stage in Hahns Peak with the hope that his prediction would be correct. Lo and behold, several hours later the stage was held up. When they opened the stage door, Tracy and Lant were looking down the barrels of a couple of shotguns. They were returned to jail at Hahns Peak. This time the two decided to sleep days in order to spend their nights hollering, screaming, and generally making all the noise they could in order to keep the townspeople awake. It worked — within days they were transferred to the Pitkin County jail at Aspen. Once again they escaped after almost beating a guard to death.

David Lant disappeared. Some say he was killed by Tracy, but there seems to be no evidence to substantiate that claim. Harry Tracy was to become even more notorious. He was jailed in Oregon — and escaped. In 1902, near Davenport, Washington, while bleeding profusely and with a posse in close pursuit, Harry Tracy shot himself to death to avoid being taken alive.

Tracy and Lant weren't the only ones to break out of jail in Hahns Peak. Tom Horn, the famous killer for hire, also escaped from jail there.

According to George Crofutt's *Grip-Sack Guide of Colorado* (1885 edition) the population of Hahns Peak was at its highest at 500. The mining activity around Hahns Peak declined in the early 1900s due to the high cost of gold extraction. Ore had to be carried out on burros or wagons because the nearest railroad, the Union Pacific, was 110 miles away. Today Hahns Peak is a peaceful little community with a handful of summer residents.

COLUMBINE

Location: 4 miles north of Hahns Peak

Nestled in a grove of aspen about four miles north of Hahns Peak, is picturesque Columbine. With the exception of a couple of cabins and a small store, most of the remaining original buildings are on a single piece of property intermingled with cabins that were built at a later date. The blend is a rustic little resort in which cabins are available for rent to the traveling public.

Columbine was established in 1897 shortly after the Royal Flush Mine was discovered just two miles to the east, on the west slope of Hahns Peak. Although it was the best mine in the vicinity, it opened and closed several times and never was a large producer.

The Master Key mine just north of Columbine was another mine which never realized much of a bonanza.

Although there were once expectations that Columbine would become a substantial city, today it is only a small stop for tourists traveling Forest Route 129.

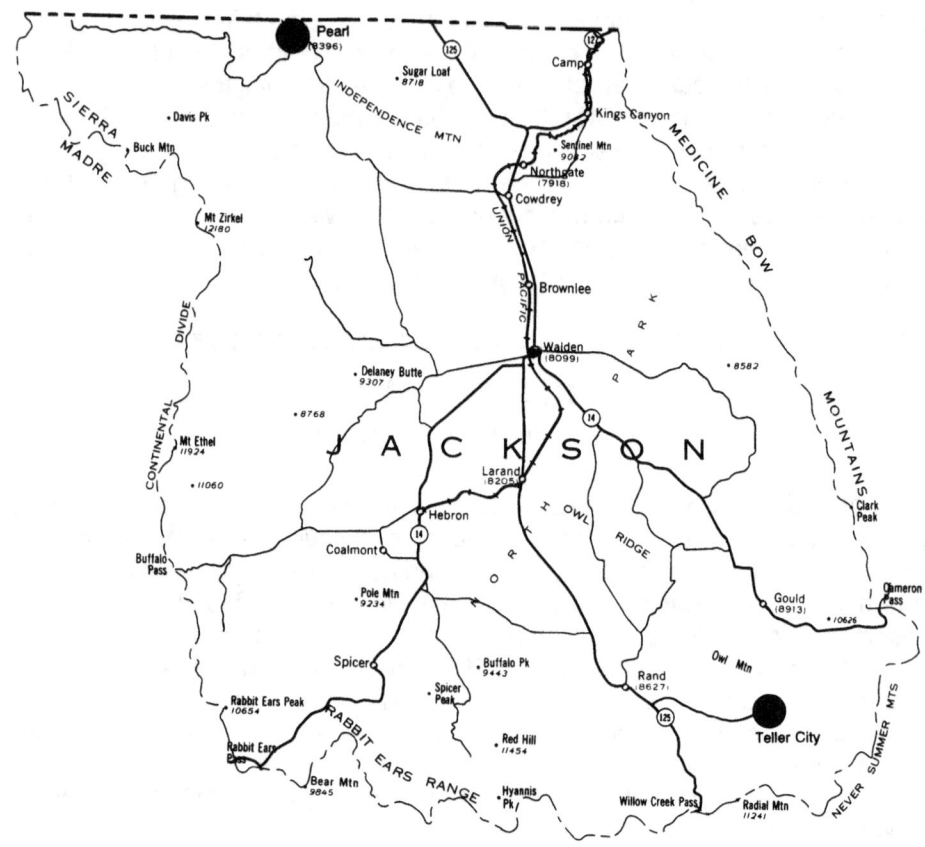

JACKSON COUNTY

"The farther it is to a place — the farther the way back."

TELLER CITY
Location: 10 miles east of Rand

Teller City is an interesting ghost town, if for no other reason, because of the vast number of log cabin remains scattered through the woods. The townsite covered a tract of 18 acres. Streets were designated by letters and numbers for simplicity. C Street was the main avenue in town. Teller City was established in 1879 after rich silver deposits were located in the vicinity. It was named in honor of Senator Henry M. Teller who was Secretary of Interior during the administration of President Chester A. Arthur. It is doubtful however that Teller ever visited the community.

During it's short life Teller City had a 40 room hotel — the Yates House (which had a piano). There were also mining and assay offices, a post office, blacksmith shop, stables, two doctors' offices, a newspaper, many stores and parlor houses, and possibly most important to the miners — 27 saloons. Population peaked in 1882 at about 1,300.

One of the best mining properties was the Indomile. It produced silver in "almost inexhaustible quantities." A pump broke, and a shaft flooded. Miners bailed water for seventy-two hours before the mine was forced to close. Shortly thereafter, the first devaluation of silver occurred in 1884. People who originally came in a hurry — left in a hurry. At an intersection in the middle of Teller City, a sign tells the story well, "The city was abandoned so quickly that dirty dishes were left on tables, and clothes were left hanging in closets."

Two illuminating stories are told about Teller City. An eastern mining company hired two men to sink a 100 foot shaft into the ground in order to determine the quality of ore at that depth. After digging fifty feet they hit water which was too difficult to bail. So they started another hole and dug it fifty feet also. Upon completion they furnished an affidavit affirming that they had completed 100 feet. After receiving payment, the two men left in a hurry never to be heard from again.

There was a fellow named Sharp who worked at the Wolverine Mine between Gaskill and Tyner. Everyone bragged that he could out-run any other man. Another chap from Grand Lake boasted that he had a man named Montgomery who could beat Sharp. A foot-race was staged on C Street in Teller City. Miners came from all around and the betting was heavy. According to the story, Sharp had a large portion of the purse in his possession as he ran down C Street. He crossed the finish line a few feet behind Montgomery — but just kept on running into the thicket. There, one of Teller City's parlor girls awaited him with a saddled horse. They galloped away and were never seen again.

PEARL

Location: 19 miles northwest of Cowdrey

Although surrounded by some beautiful country, Pearl lies in a rather desolate, treeless meadow at a fork in the road close to the Wyoming border. The remains of an old smelter nearby reflect on the mining days that once were.

Whether or not the town was named for Pearl Ann Wheeler, the original postmistress, as some believe, or whether it was named, as others speculate, for a daughter of Benjamin Franklin Burnett (who developed several Colorado towns) is uncertain. Regardless, there is not much left of the community.

The few houses which remain were either remodeled or built after the mining days. The original main street of Pearl somewhat splits the fork of the two roads which exist today.

Just a few of these buildings remain today at Pearl. *Colorado Historical Society.*

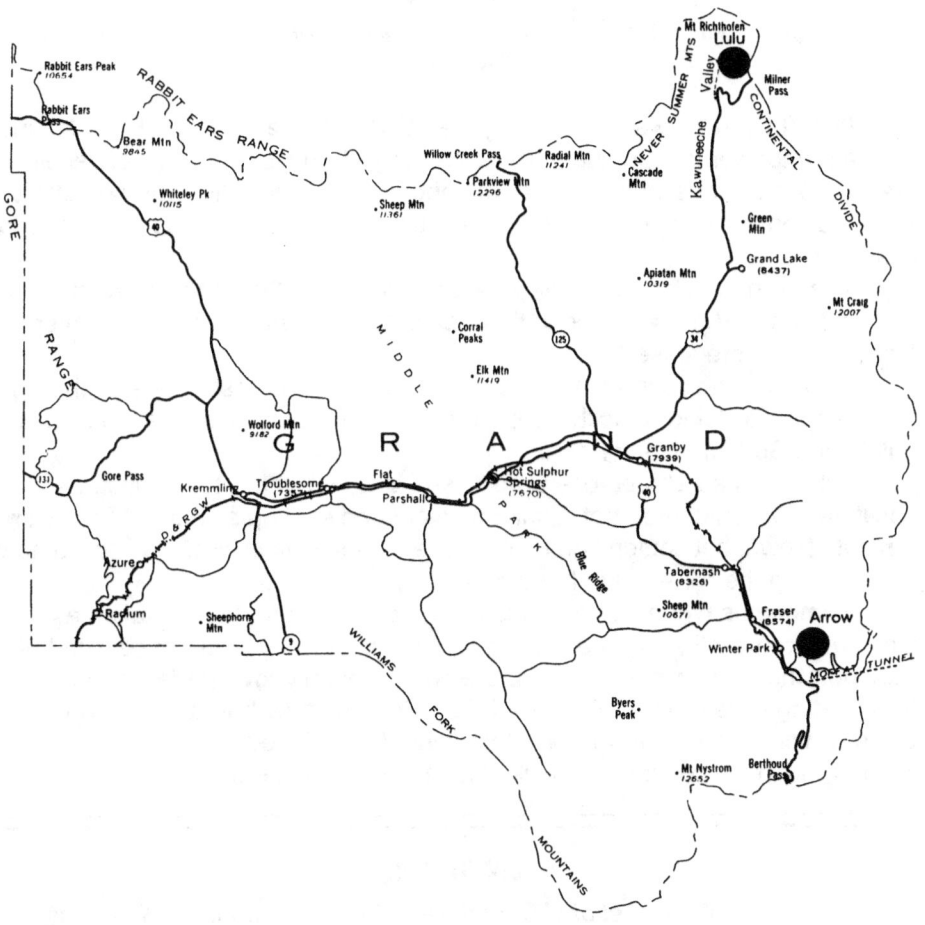

GRAND COUNTY

LULU

Location: 32 miles north of Granby via U. S. Highway 34 and the Colorado River Trail

Lulu (Lulu City) was founded by the Middle Park and Grand River Mining and Land Improvement Company. It was named for the eldest daughter of Benjamin Franklin Burnett one of its charter members. The word City was sometimes added to convey an air of importance -- but, the settlement was usually referred to simply as Lulu.

The town, which had a peak population of about 200, came in a rush -- and went in a rush. Miners flocked to the area in 1879 when silver was discovered. Many claims were staked.

Lulu was ambitiously platted. Surveyors laid out many streets -- 1st through 19th. Several buildings quickly popped-up including saloons, stores, and a fancy hotel -- the Godsmark and Parker.

Lulu even had a "suburb" -- Dutchtown -- a stone's-throw to the west. Legend has it that two Dutchmen got drunk one night and shot up a saloon. They were run out of town, but couldn't go far because of their claims. So, they built a new camp just up the hill -- out of harm's way.

Sometimes when good strikes are made on the surface, it takes a lot of digging to realize that an area is not profitable. Such was the case with Lulu. Although there were many mines, ore was generally low-grade and costs of transporting it were high. People left as quickly as they had arrived. The post office, which was established in 1880, was discontinued in 1883. The site is located within the confines of Rocky Mountain National Park.

ARROW

Location: 4 miles east of U.S. Highway 40 on the road to Rollins Pass

Restaurants in small mountain towns sometimes left a lot to be desired. Such was the case with Jack Graham's place. There were no menus. Customers paid 25 cents for a tin plate and cup. The tin plate was used for whatever food Graham happened to be serving. The cup was provided so a customer could draw from the barrel of water, if a drink was desired.

Actually, dining improved in Arrow. Adjacent to the Denver & Salt Lake Railway depot was a sign on the wall which read "Dining Room." The official name of this establishment was the Denver Railroad News and Hotel Company Eating House. Another business place with a rather descriptive name was the Furnished Rooms Establishment. The Chancey De Puy Hotel was a fancy name for a not-so-elaborate hotel. During its short life, Arrow also had a couple of other hotels, several stores, 16 saloons, a livery stable, and a "red-light" district where parlor houses and gambling dens operated side-by-side. There was never a church -- but there was a Sunday School. There was also a waterworks and several sawmills. Although Arrow had only about 200 inhabitants, the weekend population swelled considerably. Many miners and railroad workers flocked in to pick up their mail at the Arrow post office and spend a "hot" Saturday night at the "establishments."

An old-fashioned western ruckus stirred up in September 1906. Neil Ragland was the town constable. He also "owned and operated" some of the "sporting" girls in town. Indian Tom Reynolds was a half-breed who drank heavily and couldn't take a joke. One of Neil Ragland's girls hid $20 of Indian Tom's money behind a picture and wouldn't tell him where it was. The incensed and drunken breed went on a rampage. He rode his pinto into Jack Graham's place and sprayed bullets everywhere. The harassments continued the following evening until it was all Neil Ragland could take. He laid low in the shadows of the darkened Elk Saloon knowing the breed would soon arrive. When Indian Tom entered the bar -- Ragland shot him through the heart.

Although there was some mining in the area, Arrow began as a railroad construction camp. The town, which was originally called Arrowhead, sprang into existence in 1904 at an altitude of 9,584 feet. It became the first incorporated town in Grand County. The community was economically troubled from the outset. It was greatly overbuilt and the rate of business failure was high. Within two short years, most of the residents had packed up and moved on. Only the clearings remain where buildings once stood.

Arrow, on the Moffat Road below Rollins Pass. Nothing is left today but an empty site. *Colorado Historical Society.*

SOUTHWEST REGION

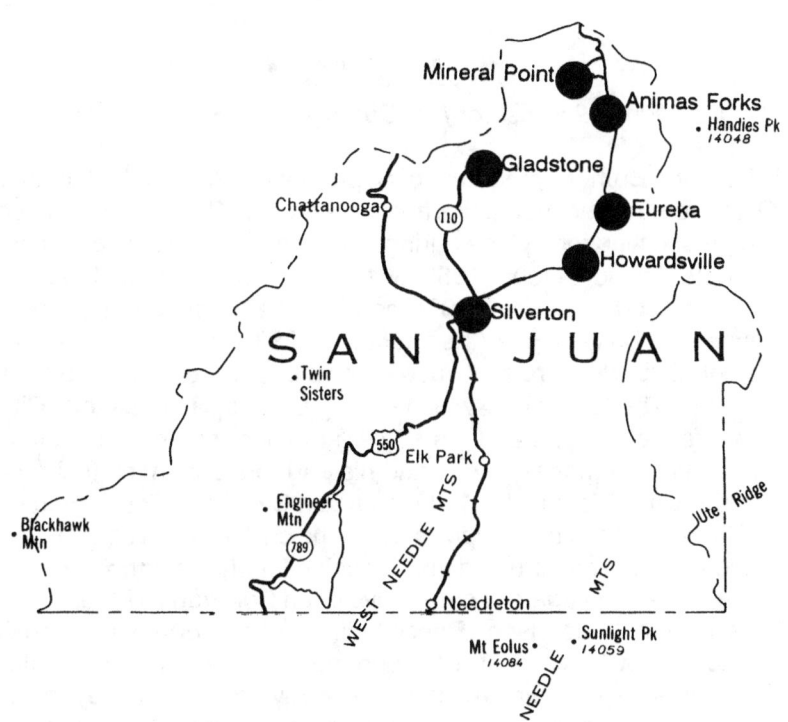

SAN JUAN COUNTY

"Considerable interest has been taken in the action of the Grand Jury, which brought in 117 indictments against lewd women ... Upon each and every prostitute, a fine of five dollars and costs were imposed."
Silverton Democrat, June 16, 1883

SILVERTON

Location: 49 miles north of Durango on U.S. Highway 550

Prior to the Brunot Agreement of September 1873, the Ute Indians, led by Chief Ouray, roamed the mountains and valleys of the San Juans. Consequently very little prospecting took place during that time. Some which did, however, is worthy of note. In mid-October 1860, on his second journey into the San Juans, Charles Baker led a party of (150?) men into the Animas Valley. Most of the group camped at Baker's Park (the future site of Silverton). According to Baker, members of his party were rewarded with values of from three to twenty-five cents per pan. The tent colony which sprang up was called Baker's City.

The threat of Indians and the Civil War minimized activity in the area for many years, until Dempsey Reese and his partners began the first productive mining at the Little Giant in 1871. Once again Baker's Park was home to new settlers. There are conflicting reports as to who built the first cabin in the vicinity. Tom Blair could have constructed the first cabin (in what later became known as Arastra Gulch) — then again, it could have been Col. Francis M. Snowden — or possibly even John P. Johnson. Reese built his cabin north of town on Cement Creek. The Brunot Agreement, which opened the area for white settlers, was actually ratified in 1874 — and the townsite of Silverton was surveyed that same year. The group which weathered in for their first winter that year included eight women. Pete Schneider delivered water in barrels to the earliest townspeople and became wealthy by selling it for fifty cents per bucket. The sheriff was also the mail carrier. Mrs. W. E. Earl led the campaign to raise funds for a schoolhouse. Upon its completion the structure also housed the first town offices and was used for church services as well. The first addition of John Curry's *La Plata Miner* rolled off the press on July 10, 1875.

Among the earliest businesses in Silverton (those established by 1875) were the general store of Greene & Company, R. C. Luesley's general merchandise store, the drugstore of B. A. Taft, the Ambold Brothers' Meat Market, and an assay office. Also, there were attorneys' offices, a doctor's office, a post office, but no jail house. The first lawbreaker was chained to the floor of a cabin. There were places to eat and drink, and there was the Briggs House — a log hotel constructed by J. L. Briggs (it was renamed the Silverton Hotel in 1876). Lodging was also available at the Centennial Hotel (later known as the Walker House). There were also two smelters and a sawmill. The late 1870s brought continued growth to Silverton. Construction throughout this period was all wood frame. The year 1880 marked the beginning of brick and stone construction, and many of the buildings which exist today were built during the ensuing period.

The east-west streets in Silverton are numbered. Those running north and south have names. The main street through the central business district is Greene Street. Most of the saloons, gambling dens, and brothels were located one block to the east on Blair Street.

Prior to the arrival of the railroad it was difficult to haul gold and silver ores from the mines in the San Juans. Pack trains could negotiate the mountain passes only during the summer, at which time it was necessary to haul in supplies for those months when Silverton was inaccessible. During the fall of 1881 the Denver & Rio Grande began laying rails from Durango to Silverton. By early 1882, the town which awaited arrival of the railroad consisted of about 300 homes, two hotels, a schoolhouse and church, two sawmills, two brick works, a few buildings related to mining, and about 75 other business establishments. The arrival of the Denver & Rio Grande into Silverton on July 13, 1882, marked the beginning of the town's greatest period of growth.

Englishman W. S. Thomson began construction in 1882 on one of Silverton's finest buildings. The elegant Grand Hotel (later renamed the Hotel Imperial, and then again changed to the Grand Imperial Hotel) was located at the corner of 12th and Greene Streets, had hotel rooms on the third floor, town offices and council rooms on the second floor, and leased space for four separate stores on the first floor (they originally housed two clothing stores and two hardware stores).

Blair Street was one of Colorado's most notorious red-light districts. It was lined with saloons, dance halls, gambling dens, elegant bordellos and seedy cribs — and it operated "wide open" twenty-four hours a day. A popular bordello was Jack Gilheany's "Laundry." According to one source, "If you went in with any money, you came out clean." The Bon Ton, Diamond Belle, Monte Carlo, and the Tremount were some of the larger houses on Blair Street. Silverton continued to serve bootleg whiskey during Prohibition, without any pressure from local law enforcement officials. Everybody pitched in to move the whiskey to safe hiding whenever word was received that Federal Revenue agents were nearby.

Like many towns in the old west, Silverton had a few gunfights. In October 1878, after an argument and fistfight between Tom Milligan and Bill Connors, Connors advised Milligan that he would shoot him the next time he saw him. Shortly thereafter as Milligan was walking down Greene Street he spotted Connors in front of the Silverton Hotel. Both men drew and Milligan shot Connors through the stomach. He died three days later. Milligan was acquitted on the grounds of self defense.

There are several variations to the Harry Cleary story, and this is one of them. On August 23, 1879, Cleary and "Mexican Joe" became very rowdy while drinking at Brown and Cort's Saloon on Greene Street. James M. (Ten Die) Brown, one of the saloon's owners, escorted Cleary out the front door. Cleary turned and shot Brown. Brown was able to get off some shots, and one stray bullet hit handicapped night watchman Hiram Ward in the left shoulder. Ward, in turn, also shot Brown. Although it was more likely one of Ward's bullets that penetrated Brown's heart, Cleary was blamed for the killing, arrested and jailed.

Late that night a mob dragged Cleary from the jail house and lynched him behind the blacksmith shop.

At 11 p.m. on August 24, 1881 La Plata County Sheriff, Luke Hunter, arrived in Silverton with warrants for the arrest of members of the Stockton-Eskridge gang. Burt Wilkinson, Dyson Eskridge, and Kid Thomas (a black man who was also known as the "Copper Colored Kid") had been drinking at the Diamond Saloon. Sheriff Hunter rounded up Silverton's marshal, D. C. (Clate) Ogsbury. As the two men walked down Greene Street, Wilkinson and Eskridge opened fire killing Ogsbury instantly. The two escaped on foot while Kid Thomas, who apparently fired no shots, rounded up their horses. He was apprehended near the stable and carried to jail. Once again, a mob took the law into its own hands as they dragged the black lad from jail and lynched him behind the old county building. Burt Wilkinson was turned in for a $2,500 reward by the gang leader Ike Stockton. Once again, the mob ruled. According to the *San Juan Herald*:

> " ... a party of masked men suddenly appeared before the guards at the jail and overpowered both of them and the jailer, went into the jail and seizing Wilkinson, passed the noose about his neck and asked him if he had anything to say before his death. He replied: 'Nothing, gentlemen, Adios!' He was perfectly composed to the very last, got up on a chair and assisted the vigilantes to hasten the hanging."

About a month later, Ike Stockton was shot in Durango by Deputy Sheriff Jim Sullivan. The bullet shattered Stockton's knee. Following amputation of his leg, Stockton bled to death.

There are many stories to tell about Silverton — and its many colorful residents. There were the earlier girls of Blair Street like Molly Foley and Blanche DeVille. Some of the prostitutes after the turn of the century were even more colorful, such as "Diamond Tooth" Leona, "Jew" Fanny, "Nigger" Lola, and "Tar Baby" Brown. Ruthless town marshal Tom Cain was in the news consistently for his questionable activities. The famous Wyatt Earp ran the gaming rooms for a while at George Brower's fancy saloon and gambling hall, the Arlington. The famed Dodge City Cow Boy Band (the original spelling of cowboy was two words) of C. M. Beeson moved its headquarters to Silverton in June 1890, and operated from town for a few years until Buffalo Bill purchased the group which became part of his Wild West Show. Part of the Guggenheim fortune was reaped in the Silverton area. Multi-millionaire Thomas Walsh of Camp Bird fame had mining interests in the region as well.

Mining was an important part of Silverton's economy. Some of the region's top mines were the Silver Lake, the Iowa and Tiger at Arastra Gulch, North Star, the famed Sunnyside (see Gladstone for discussion on the Sunnyside and American Tunnel), the Shenandoah, and the Dives. The latter two were later combined and operated by the Shenandoah-Dives Company, which also operated the North Star, the Terrible, and other properties. It also constructed a large mill in 1933.

The tourist industry is an important part of today's economy. The old coal-fired, steam-operated train runs daily during the season between Durango and Silverton. Since 1981 it has been operated by the Durango & Silverton Narrow Gauge Railroad Company (D&SNG) Additionally, there are many century-old buildings and other relics which reflect the old west that once was.

"If you have a mountain to climb, waitin' won't make it smaller."

MINERAL POINT

Location: 8 miles southeast of Ouray

Prospectors Abe Burrows, who discovered the Burrows Mine, and Charles McIntyre founded the camp at Mineral Point in 1873. It was named for the large composite of quartz and other minerals located at the site. The rich vein of the Mastodan Mine extended through Mineral Point and on for several miles.

It is said the Old Lout Mine was thought to be worthless and in the process of being abandoned when one of the miners took one last shot and uncovered a rich body of ore. The mine yielded $86,000 the first month. There were other good mines in the area including the Red Cloud, the Vermillion, and the Bill Young. The San Juan Chief Mill was constructed above the city.

Mineral Point was located high in the mountains, below the crest of Engineer Pass, at an altitude of 11,474 feet. Winters were severe — even summer nights were cold. Its summer population was approximately 200. Stage service came from each side of the range. Silver ore was also shipped out in both directions for reduction — down to Lake City in the East and usually to Animas Forks on the western side. The camp had a hotel (which housed a saloon), a general store, several restaurants, and a saw mill.

The demonetization of silver caused by the repeal of the Sherman Silver Purchase Act in 1893, spelled doom for Mineral Point.

HOWARDSVILLE
Location: 4 miles northeast of Silverton

Howardsville was named for George W. Howard, who built the first permanent cabin in the community. By 1874 there were many homes, stores, and saloons, as well as a reduction works. At that time, the town was a part of La Plata County which included the area now covered by Ouray, San Juan, and parts of San Miguel and Dolores Counties. Howardsville became the first county seat selected in western Colorado. A few short months later, it was dumped as county seat. When La Plata County was divided, so were the records. The records for that portion which was to become San Juan County were moved to Silverton, the new county seat.

The area had good mines leading down from Stoney Pass, originally one of the main entrances into the San Juans, on to Howardsville at the head of Cunningham Gulch. The post office closed in 1939, but a few residents still live up in the Gulch.

EUREKA
Location: 7 miles northeast of Silverton via State Highway 110 (east)

Alongside the Animas River, from the flat where the townsite of Eureka once existed, the foundations of the famed Sunnyside Mill (once one of the largest in Colorado) stair-step up the mountainside. The history of Eureka parallelled the history of the Sunnyside.

The Sunnyside Mine was located in 1873. At first it was a property of poor to average yield, and expensive to operate. Operating in the red, John Terry sold the property to a New York group for $300,000. After paying Terry a $75,000 down payment, they shortly became disenchanted with the mine and offered it back to Terry. With money in his pocket, Terry resumed operations. Before long the mine was producing very rich ore. The Sunnyside made Terry a millionaire. The property was worked continuously until it closed in 1931. It was renovated and reopened again — twice, closing again during World War II. The Sunnyside employed as many as five hundred persons during its most productive period. The American Tunnel (see Gladstone) was extended to tap the Sunnyside, and the mine continued to produce through most of the 1960s and '70s — in fact it has produced over six million tons of ore during its history. The Sunnyside Mine lease was purchased in 1985 by Canadian-owned Echo Bay Mines.

The area in and around Eureka had many mines and several mills. During the boom-days two thousand people lived in the community and most of them worked at the mining properties.

Otto Mears' Silverton-Northern Railroad arrived in Eureka in 1896. The railroad operated for many years, until the first closing of the Sunnyside.

At first Eureka only had one hotel, but after the turn of the century each of the large mills had its own boardinghouse. The business district was filled with the usual stores, saloons and restaurants. Eureka had a monthly newspaper — the *San Juan Expositor* — which was published by Theodore Comstock who later went on to found the School of Geology at the University of Arizona.

Eureka's history is marked by casualties and property damage from rockslides and snow-slides. Buildings at one edge of town and some of the mining properties were perilously close to the base of barren mountain slopes. The aforementioned foundations and a couple of structures are all that is left.

After visiting Animas Forks, George A. Crofutt, in his Crofutt's Grip-Sack Guide of Colorado stated, "It is a wild and rugged country, where nothing but rich mines would ever induce a human being to live longer than absolutely necessary."

ANIMAS FORKS
Location: 12 miles northeast of Silverton via State Highway 110

The town of Animas Forks was poorly platted in 1877, about two years after the first claims were staked. Those inhabitants who braved the harsh winters did so defensively, in the wake of snowstorms and dangerous drifts. As an enticement to live near timberline, settlers were offered free lots and construction assistance. Many responded, and the community blossomed.

The town was built with an air of permanence. Buildings were well-constructed with finished lumber and shingled roofs. False-fronted buildings lined the main street of town. By the mid-80s, Animas Forks had three hotels (the Mercer, Flagstaff, and Kalamazoo House), two assay offices, many stores and saloons, a telephone system, and a newspaper — the *Animas Forks Pioneer*. The jailhouse, which contained two cells, was constructed entirely of two-by-sixes laid flat. At the south edge of town stood the huge Gold Prince Mill.

The boardinghouse of Mrs. Eckard was very popular with miners in the Animas Forks' area — and so was she. Once a freeloader absconded without paying three months' board. A "posse" of Mrs. Eckard's friends found the scalawag in Silverton and offered him a choice — pay up or be hanged. Mrs. Eckard received her money, and no one ever cheated her again.

The Gold Prince was the most productive mine in the area. Other valuable mining properties included the Early Bird, Columbia, Silver Coin, Red Cloud, Eclipse, Little Arthur, and the Iron Cap.

Near the turn of the century, famed pathfinder Otto Mears built the Silverton Northern Railroad spur from Silverton, through Eureka, to Animas Forks.

Amidst the ruins of Animas Forks stands a once stately and elegant

Victorian home. A prominent bay window faces toward the Animas River. Locals say it belonged to Thomas F. Walsh, discoverer of the famed Camp Bird Mine. One legend says, his daughter Evalyn Walsh McLean (once an owner of the Hope Diamond) was born here. Another indicates that she lived here while writing her biography, *Father Struck It Rich*. Evalyn was born in Denver and it is unlikely she ever lived in Animas Forks.

Although some mining continued, Animas Forks was a ghost-town by 1923. The railroad bed once again became a road when the tracks were removed in 1942.

GLADSTONE

Location: 7 miles north of Silverton via State Highway 110 (west)

A chlorination works was constructed on the road up Cement Creek from Silverton to Poughkeepsie — a short-lived mining camp a few miles above. Gladstone began as a few employees established "residency" near the works.

The Sampson Mine was located in 1882. Other discoveries followed. The story of Gladstone, however, is centered around the Gold King Mine. In 1887, Olaf Nelson, discovered a good vein while working the Sampson. With the thought that there might be more rich ore nearby, he staked out an adjacent claim. Nelson worked his claim on a part time basis for three years until his death. His widow sold the Gold King in 1894 to Cyrus Davis and Henry Soule for the sum of $15,000. The new owners invested more into the property and the expansion paid off. The mine yielded over one million dollars during the next three years. Much of its success was due to Willis Kinney, who managed the mine and originally suggested its purchase.

The prosperity of Gladstone paralleled that of the Gold King. Rows of white company owned cottages sprang up to house employees. Gladstone, which was named for the Prime Minister of Great Britain, had boardinghouses, restaurants, saloons, and dancehalls. There was a newspaper, the *Gladstone Kibosh*. A smelter was constructed. The arrival in 1899 of Otto Mears' Silverton, Gladstone and Northerly Railroad, a narrow gauge, was cause for much celebration.

Tragedy struck the Gold King Mine in 1907 when fire trapped three miners. Two of the three, and four rescuers, died as a result of smoke inhalation. Further trouble beset the mine when litigation between heirs and stockholders closed the property in 1910. The mine re-opened in 1918, again under the management of Kinney.

The American Tunnel Project of Standard Metals was started in 1959 when construction began to enlarge and extend the Gold King Tunnel (already one mile long) in order to tap the lower extensions of the Sunnyside Mine (also see Eureka). An average of 600 tons of ore were processed per day through the 1960s. Since then production has been on an off again-on again basis. On Sunday, June 4, 1978 disaster struck. The spur vein of the Sunnyside was being mined seventy feet below Lake Emma which broke through dumping thousands of gallons of water and a million tons of mud into the mine — a mess which required two years to clean up. Luckily nobody was killed as the entire crew of 125 workers had the day off when the break-through occurred. The project closed in 1991.

Burros on Silverton's 13th Street are loaded with track rails for the mines. *Denver Public Library, Western History Department.*

The Sunnyside Mill stair-steps up the mountainside at Eureka. *Colorado Historical Society.*

The town of Animas Forks is located near timberline. *Collection of Dave Southworth.*

Gladstone and the Gold King Mill. *Colorado Historical Society.*

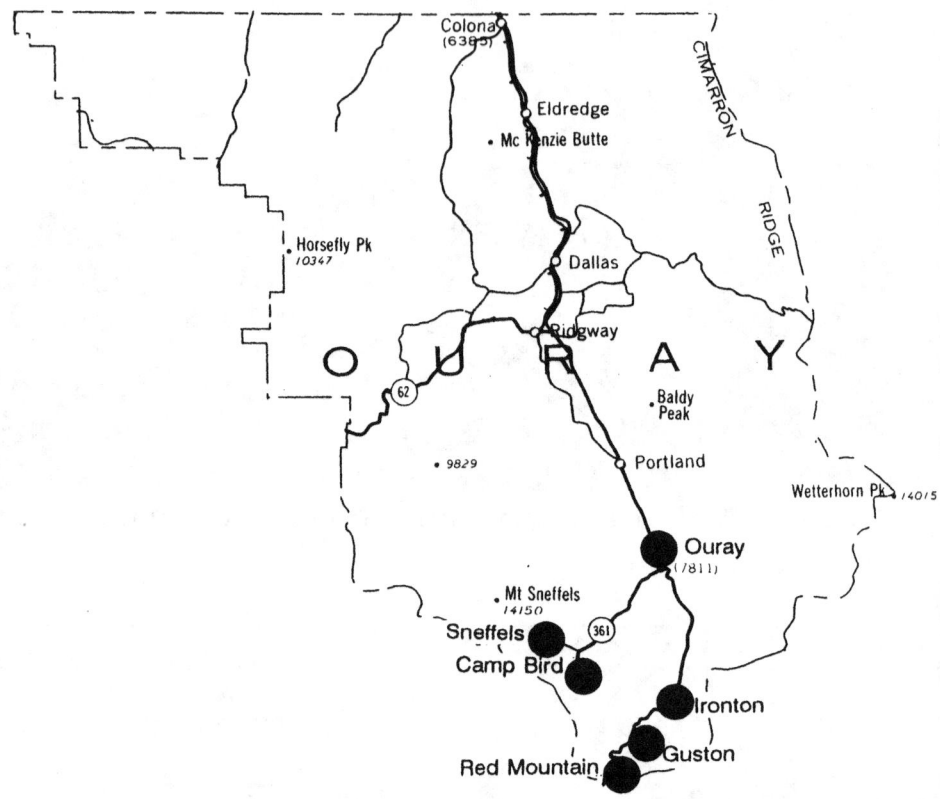

OURAY COUNTY

Upon the death of a Ouray prostitute, David F. Day editor of the Solid Muldoon wrote the following obituary:

"Charlotte,
Born a virgin, died a harlot.
For 14 years she kept her virginity,
An all-time record for this vicinity."

OURAY

Location: 37 miles south of Montrose, or 23 miles north of Silverton, on U. S. Highway 550

Ouray is located at the bottom of a bowl of steep cliffs which drop below the jagged, snow-capped peaks of the Uncompahgre Mountains. The valley is stark — yet spectacular and beautiful.

In 1875 A. J. Staley and Logan Whitlock, while fishing along the Uncompahgre River at the mouth of Canyon Creek, discovered the Trout and Fisherman lodes. During the same year, A. W. Begole and Jack Eckles located the Cedar, Clipper, and Mineral Farm lodes — and the rush was on.

A town was platted which was first called Uncompahgre City, then renamed Ouray. Through his unparalleled strength of character and artful negotiations, Chief Ouray of the Utes was more responsible than any other person for gaining peace between the Indian and white man in Colorado. It is fitting that the city of Ouray is named in his honor.

Printing presses were carried over the mountains by wagon train, and the first newspaper was established in 1877 — the *Ouray Times* (later named the *Plaindealer*). Shortly thereafter a second newspaper began operations — the *San Juan Sentinel*. The third, and most famous newspaper appeared in September 1879. The opinionated and controversial *Solid Muldoon* of David Frakes Day was a widely quoted publication. By this same year the population of Ouray was approaching 1,000. There were three hotels, three churches, several stores, saloons, gambling dens, and a "red light" district which was located at the north edge of town.

W. J. Benton's Star Saloon was the first frame building constructed in town. When Ouray became county seat, the saloon was converted into the courthouse. The bar was removed and the first floor became the city hall. The rooms on the second floor, which were once sleeping quarters, were changed into county offices.

The only known lynching of a woman in Colorado mining camp history occurred in Ouray. In 1884, after a man and woman had been jailed for allegedly raping and killing a 10 year old orphan girl, an irate mob of citizens dragged the couple from jail and hanged them.

Mrs. Dixon operated one of the earliest boardinghouses. In actuality it was a cabin without much room. Guests brought their own bedrolls and slept under the tables where Mrs. Dixon served meals. She did well and saved her money, then built a fine hotel — the Dixon House.

Transportation was a major problem during the early years, and remained so until the arrival of master road builder Otto Mears in 1881. Increased access boosted Ouray and its economy. Within a few years the railroad arrived (in 1887), advancing the community even more.

Snowslides and bitter winters were also a problem during the early history of the mining district. On one occasion a mining accident injured two workers at the Terrible. While enroute to the location of the accident, four rescuers were killed in an avalanche. Seven miners and twenty-seven horses and mules lost their lives in another slide. Later, the disastrous blizzards during the winter of 1905-06 claimed many lives. On the lighter side, a story is told of a man who was buried in a mine by a snowdrift. He tunneled his way out and walked into Ouray two days later, only to find his name in the obituaries.

In addition to the aforementioned mines, some of the better producers were the American Nettie, the Virginius, Chief Ouray, the Bachelor, Wedge, Khedive, Calliope, and the Banner American. The silver crash of 1893 made a dent in the economy, but Ouray wouldn't crumble. Whereas the highest percentage of mining had been silver, the district turned its attention to gold and other minerals.

The Bachelor Mine is an example of how faith, hard work, and perseverance paid off. The owners were a cook, a mail carrier, and a prospector. They put all of their earnings back into the mine. After tunneling 700 feet they cut a gold vein and their prayers were answered. The Bachelor Mine produced about $30,000 per month.

By the 1890s much of downtown Ouray had been rebuilt in stone and brick — such as the elegant Beaumont Hotel, at the corner of Main and Fifth Streets. Its stately architecture and luxurious interior made it the finest hostelry in the Uncompahgres. The railroad brought more people, so consequently there were more gambling dens and parlor houses. The number of saloons had increased to thirty. In 1890 the population of Ouray County was 6,510. Following the repeal of the Sherman Silver Purchase Act, the population dwindled to 4,731 in 1900. Generally, there was a continuing decline until 1940 when the population had dropped to 2,089, at which point it stabilized. Fifty years later, in 1990, the population was 2,295. Picturesque Ouray has continued to be a popular spot for summer tourists.

"Before you climb into a gulch, know how you'll get out."

SNEFFELS

Location: 6 miles southwest of Ouray via State Highway 361

The Wheel of Fortune, Yankee Boy, and Virginius mines were among the earliest discoveries near Sneffels — and the little camp became a lively place. Additional discoveries followed, such as the Ruby Trust, Hidden Treasure, Humboldt, Governor, and the Senator. A fine producer was the Atlas, which like several others, had a mill and boardinghouse. The mines of Mount Sneffels yielded an estimated $27,000,000 during the boom years. Some of the ore assayed as high as $40,000 per ton. Mount Sneffels (as well as the settlement) was named for Jules Verne's Icelandic peak in *A Journey to the Center of the Earth*.

Sneffels was both a gold and silver camp. The silver came easy, but gold was only encountered after shafts were driven very deep into the mountain side. In 1884 the Revenue Tunnel was constructed to intercept the rich Virginius. The bore was made at a point nearly 3,000 feet in elevation below the mouth of the original shaft then extended about three miles into the mountain. The project, which was financed by the Thatcher brothers of Ouray, cost $600,000 but paid for itself almost immediately. The tunnel not only solved water and ventilation problems but opened up enough new ore to keep the mills running for years to come. The property became known as the Revenue-Virginius, employed 600, and developed into one of the best mines in Colorado.

The road from Ouray to Sneffels is steep, narrow in places, and partially a shelf notched in the mountain side. At one time the road was a continuous traffic jam, with packed trains, freight wagons, loggers, stagecoaches, and individuals walking or on horseback. The road was treacherous during winters and the result was many accidents. There is much less traffic today. Along it's route, and from Mount Sneffels above, the scenery is spectacular.

Indians called the train, "Plenty wagon — no horse."

RED MOUNTAIN

Location: 13 miles south of Ouray on U.S. Highway 550

A few claims were staked in the vicinity of Red Mountain as early as 1879, but nothing much happened until late 1882. A large cavern was discovered which glittered from the rich lead carbonate lining the walls. The find was dubbed the National Belle. For fear they would miss out if they waited until spring, prospectors flocked to Red Mountain in the dead of winter.

The ground was frozen, and proper foundations couldn't be built.

Nonetheless, several mining camps popped up before the snow thawed, and others followed on both sides of Red Mountain Pass. The community now generally called Red Mountain was once known as Red Mountain Town, and was the largest of the lucrative camps in the Red Mountain District. Guston and Ironton were substantial towns to the north. Additionally, Yankee Girl was a small camp adjacent to the Yankee Girl Mine near Guston. South of the summit were short-lived Red Mountain City and Sheridan Junction (Chattanooga).

Red Mountain (Red Mountain Town) became the largest and the most prosperous in the district. The camp which began in January 1883, was a boom town in two short months. In March 1883 the *Solid Muldoon* (Ouray) stated: "Five weeks ago the site where Red Mountain now stands was woodland mesa, covered with heavy spruce timber. Today, hotels, printing offices, groceries, meat markets, ... a telephone office, saloons, dance houses are up and booming; the blast is heard on every side and prospectors can be seen snowshoeing in every direction." Initially, Red Mountain was accessible from several directions — and all were dangerous trails. The difficult access was solved by roadbuilder Otto Mears. After the "Grandest Highway in the Rockies" was surveyed, Red Mountain residents began to scramble for lots in the flat close to the highway site. Mears' road — generally the predecessor of the Million Dollar Highway — was constructed north from Silverton through Red Mountain and on to Ouray. The whole town moved about a half-mile in order to be close to the toll road.

The road was not enough, however, so Otto Mears built a railroad which connected with the Denver & Rio Grande at Silverton then snaked through the mountains to the north for twenty miles. The "Rainbow Route" as it was dubbed, arrived into Red Mountain in September 1888 amid much fan-fare.

During the boom years saloons and gambling halls never closed — day or night. Violence was commonplace. A Saturday rarely passed that didn't end up in some kind of brawl. The lively community also had other forms of entertainment. Dances, plays, concerts, and sporting events were held on a regular basis, and were attended by guests from as far away as Ouray and Silverton.

In addition to the aforementioned National Belle Mine, there were many other top producers. The Yankee Girl (see Guston) topped them all — producing over $8,000,000. Three of the earliest finds were the Congress, the Summit, and the Enterprise. Population estimates during the boom years range from 1,000 to about 10,000. It is more likely that the latter figure included the whole mining district, with about 3,000 in the town of Red Mountain. Once the prospecting frenzy passed and everyone settled in to work the established mines, the population was closer to the 1,000 estimate. Following the demonetization of silver in 1893, the population dropped to 400 and continued to decline. The population by 1896 was 40. Off and on again mining continued through the ensuing years, but all of the camps in the Red Mountain Mining District died long ago.

"Ironton is the first real ghost town I ever saw. Many of the original buildings were intact in 1951 during my initial visit (at age 13). Through the '50s buildings weathered badly — and some disappeared. By the mid-'60s there was still an abundance of old artifacts, signs, etc. — even the chassis of an old Model T. Then came the barnwood craze in the '70s and much was stripped. Following more deterioration, the '90s leaves us with just a few buildings still standing — with flapping tin roof panels and creaking floors — to remind us of the Ironton that once was. This site, possibly more than any other, spurred my interest in Colorado mining towns."

— Dave Southworth

IRONTON

Location: 8 miles south of Ouray on U.S. Highway 550

Ironton is another town that blossomed during the Red Mountain mining craze. The town began in January 1883 when snow was cleared and 300 tents pitched. The camp initially had two names — Ironton and Copper Glen, the latter which was quickly dropped.

The American Girl, Cora Belle, Mountain King, Silver Bell and the Lost Day were all mines that produced well. Much of the economy and livelihood of the community, however, was due to the enormous success of the Yankee Girl Mine at nearby Guston. Even though Ironton had ten saloons, it grew with an element of refinement and class. Merchants of some of the better stores in Ouray and Silverton opened branches at Ironton. People from Red Mountain and Guston often shopped at Ironton's specialty shops. The town emerged as a supply center for the region. It also was important as a transportation center and stopover for stagecoach lines (such as the Ouray Stage and Bus Co.) and supply wagons.

Arrival of the railroad into Ironton boasted activity even more. A grand celebration welcomed the first train in November 1888. The rails of Otto Mears' Rainbow Route left Silverton and climbed across Red Mountain Pass at an elevation of 11,650 feet, then descended to Ironton. Upwards of 20,000 tons of ore were transported out of the Red Mountain District by train annually.

Ironton never had a major conflagration — possibly because it was well equipped. There was a firehouse and hose company, and fire hydrants were scattered throughout the town. Not only were the waterworks impressive, but Ironton also had an electric light plant.

The silver crash of 1893 closed most of Ironton's mines. People moved on in search of greener pastures. A new, but brief, boom occurred when gold was discovered in 1898. Ironton remained inhabited into the early 1930s. Outside of Ironton proper stood a cabin nestled in a nearby grove of aspens where a recluse named Larson lived on for another three decades.

"You can't be hurt by the words you don't say."

GUSTON

Location: 11 miles south of Ouray on U.S. Highway 550

John Robinson had much to do with four of the earliest discoveries in the vicinity. During the summer of 1881, Robinson and his group of prospectors staked the Guston claim. At the time, it yielded low-grade ore and the project was nearly abandoned. The following year, he discovered nearly pure galena about 300 yards from the Guston and staked the Yankee Girl. Further digging at the Yankee Girl didn't yield much either, and Robinson's operating capital was running low. He knew the mine had promise, however, and staked claims on opposite sides which he named the Robinson and Orphan Boy. He then sold the Yankee Girl to a partnership for $125,000 providing the capital he needed. The Yankee Girl sputtered for its new owners as well — but when it hit, it was a bonanza. Its production exceeded 25% of the entire district — yielding over $8,000,000. The Guston, Robinson and Orphan Boy were also successful endeavors. Other good mines at Guston were the Saratoga, Candice, Paymaster and the Genesee-Vanderbilt.

Guston was a small town, with a population of about 300, yet it had the distinction of having the only church in the entire Red Mountain Mining District. In 1891, the Rev. William Davis was sent by the Congregational Church to Red Mountain in order to establish a mission. The pastor received a polite but cool reception. Finding no place in which to conduct worship services, he went to Guston where he was cordially received. More determined than ever, Davis set out to raise money for construction of a church. Contributions were received from Silverton to Ouray, the land and pews were donated, and the little church became a reality. Somebody suggested that a mine whistle should be installed in the belfry in order to be heard near and far. And so it was. It is the only church known to have announced its services with the shrill blast of a whistle.

In 1899 a stagecoach between Camp Bird and Ouray was robbed by two masked men who fleeced the passengers but overlooked a shipment of gold worth $12,000.

CAMP BIRD

Location: 5 miles southwest of Ouray via State Highway 361

Flamboyant Evalyn Walsh McLean wrote a book entitled *Father Struck It Rich*, in which she tells the story of Thomas Walsh and the Camp Bird Mine.

Walsh was an Irishman and a carpenter by trade. Bridge building brought him to Colorado. His burning desire to get rich quick turned him to prospecting. In 1896 Walsh discovered gold — very rich gold — in Imogene Basin high above Ouray. He immediately purchased approximately 100 claims in the area and consolidated them under the name of Camp Bird. It wasn't long before his mining property produced over $1,000,000 per year, and it eventually became the second largest producer in Colorado (only the Portland Mine near Cripple Creek was larger).

When Evalyn married Edward B. McLean, whose family owned the *Washington Post*, the newlyweds received $100,000 from each family as a wedding gift. Later Evalyn purchased the famous Hope Diamond which she dangled in front of Washington and Denver society.

For his employees Walsh built a posh boardinghouses. The facilities which accommodated 400 men had marble-topped lavatories, electric lights, steam heat, china plates, and even a piano. Meals were often served that rivaled the finest restaurants.

Walsh was already a multi-millionaire when he sold the Camp Bird properties, in May 1902, to an English syndicate for 3.5 million dollars in cash, a half-million in shares of stock, and royalties on future profits. To show his gratitude, the generous Thomas Walsh gave employees bonus checks of up to $5,000. Before his death in 1910, he had received 6 million dollars from the sale of the Camp Bird.

Each winter snow was always a problem for the high-mountain community. Through the years snowslides killed several men. Geographically, the site is reached via narrow, winding mountain roads. It was even necessary to construct a two-mile long aerial tramway from the mines to the mill. Camp Bird produced well however — and continued to produce.

Ouray about 1886. The Beaumont Hotel (at right center) is under construction. *Colorado Historical Society.*

Sneffels at the turn of the century. *Denver Public Library, Western History Department.*

OURAY COUNTY 247

Red Mountain Town at elevation 11,300.
Denver Public Library, Western History Department.

The "Rainbow Route" of Otto Mears arrives at Red Mountain Town in September 1888. *Denver Public Library, Western History Department.*

Loading up supplies at Ironton. *Denver Public Library, Western History Department.*

The aerial tram at Camp Bird carried ore and miners. *Colorado Historical Society.*

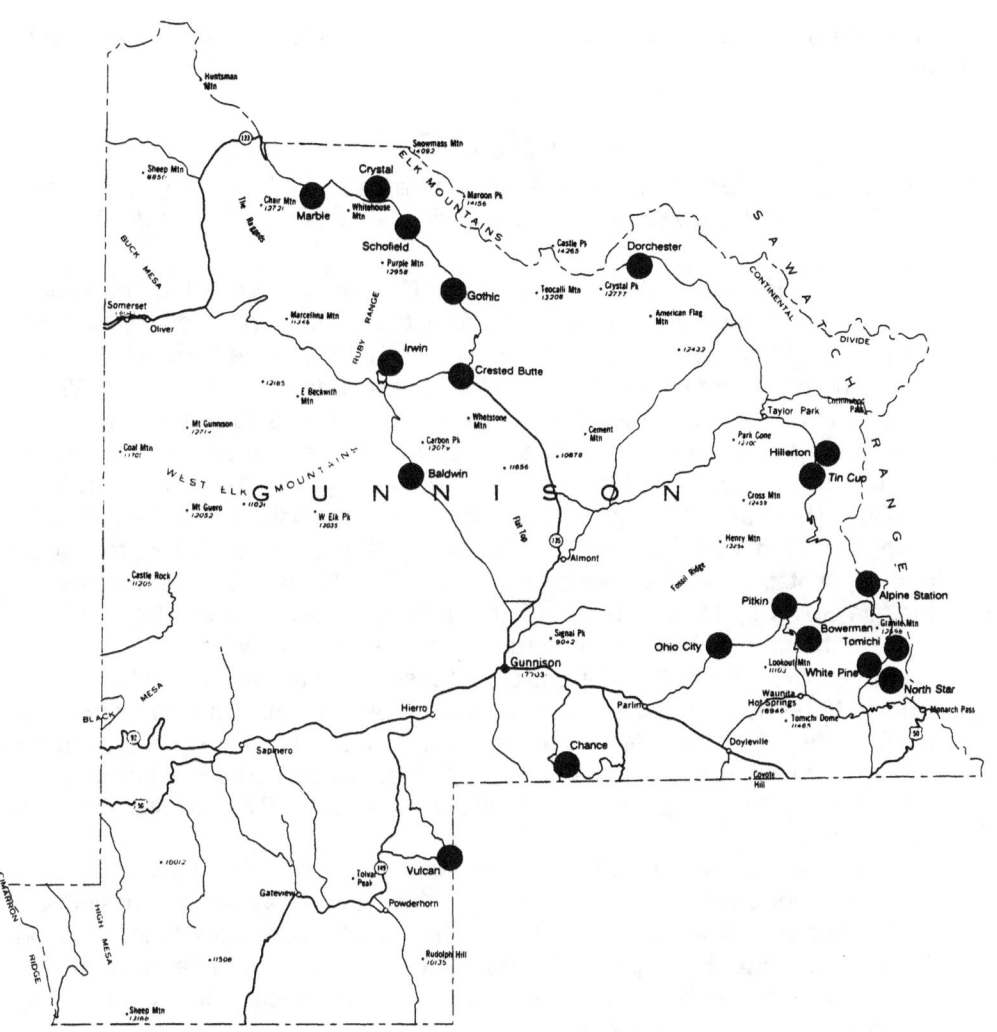

GUNNISON COUNTY

"Tin cups and tin plates last longer 'cause they're easier to fix after a boisterous meal."

TIN CUP

Location: 43 miles northeast of Gunnison via State Highway 135, and Forest Routes 742 and 765; or 26 miles north of Pitkin via Forest Route 765

Jim Taylor, Gus Lamb, Ben Gray and Charlie Gray were the first known prospectors in the vicinity of what was to become Tin Cup. They crossed the Continental Divide and pitched camp. That night their horses broke loose and wandered off. The following morning they followed the horse tracks to Willow Creek and found their horses. Thirsty, Ben Gray knelt beside a stream to scoop a drink of water. The creek bed looked promising so he scooped up some gravel in his tin cup — and there it was — traces of gold. They named the spot Tin Cup. There are several versions of this story — with the tin cup belonging to several different people. It wasn't until nearly 20 years later that a prospector named Captain Hall, who was grubstaked by Carl Hord of Denver, discovered the rich Gold Cup Mine in 1878. After that things really started to happen.

At the time Sol Bloom constructed the first permanent cabin at Tin Cup Camp in 1877, the settlement basically consisted of a few prospectors' tents. During 1879 the town experienced its most rapid growth. Tin Cup Camp was incorporated in 1880 under the name Virginia City. Because of the confusion with other communities named Virginia City, the U.S. Post Office was instrumental in convincing the townspeople to change the name. In 1882 it was reincorporated as Tin Cup.

Tin Cup had several hotels and boardinghouses of which the most notable were the Pacific Hotel, the Tin Cup Hotel, and Aunt Kate's where possibly the best meals in town were served. Kate Fisher was a former slave who arrived in Tin Cup in 1880. She operated a boarding house and her fine cooking was renown throughout the mining community. She was successful, and loved by the townspeople until her death in 1902.

Additionally, the community had two banks (one of which was the Bank of Tin Cup), a general supply and grocery, meat market, boot and saddle shop, livery stable, schoolhouse, post office, jailhouse, general mercantile store, wagon and carriage repair shop, bakery, telegraph office, Dr. John McGowan's office, and a Masonic Hall. Fire plugs were installed in 1891. The Town Hall was not constructed until 1906. Population estimates for Tin Cup have been exaggerated. During the boom years of 1880-82 there may have been as many as 5,000 in the entire mining district. Many of those were tent-dwellers who moved on. At that time the population of Tin Cup was about 1,200 — and dropped to 500 by 1884. It experienced a second surge during the flurry of new mining activity in 1903-04.

Henry Olney moved his press from Hillerton in 1881 and established the *Tin Cup Record*. There is some confusion as to how many newspapers Tin Cup had, but it had at least two, plus a few name changes.

Tin Cup was one of the wildest towns in Colorado — and had a reputation for being so. During the boom years, 26 saloons and gambling houses operated night and day. Parlor girls were quick to fleece a miner of his gold dust. Shootings and drunkenness were commonplace.

Owners of the saloons and gambling halls "controlled" Tin Cup. In 1880 a marshal was hired to give the town an appearance of being orderly. He was informed that the first person he arrested would be his last. The second marshal was also a pawn. He did little more than round up drunks, disarm them, then released them. By 1882 Tin Cup had become so lively that tough Harry Rivers was hired as marshal and instructed to maintain law and order. Rivers pushed his weight around, especially when he had too much to drink. One night Rivers had been drinking heavily and decided to badger saloon owner Charley LaTourette. He followed LaTourette home, shooting off several rounds in the process. When LaTourette reached his gate he turned, drew and fired killing Harry Rivers. LaTourette was tried and "exonerated". Rivers' successor as marshal went insane. The fifth marshal was shot and killed.

During the lively boom years of the early 1880s parlor houses were located on the south end of Grande Avenue. As mining activity declined, many of the girls moved from Tin Cup. Those who remained moved into the alley behind Washington Avenue and became known locally as the "alley girls". A story is told about "Oh Be Joyful", one of Deadwood Sal's girls. Sal was a madam who kept a very tight rein on her girls. "Oh Be Joyful" fell in love and wanted to get married, but Sal wouldn't release her from her contract. Late one night the firehouse ladder mysteriously disappeared. It was found the next morning propped at "Oh Be Joyful's" window. She had been whisked away to a quick wedding and points beyond.

Tin Cup has never had a church. On those occasions when an itinerant preacher came to town, services were held in the schoolhouse. Tin Cup was such a rowdy place that no preacher was ever encouraged to establish a permanent church. When priests or ministers were unavailable, weddings and funerals were usually conducted by Harry R. Morris who at one time or another was Police Magistrate, Clerk and Recorder, Postmaster, Justice of the Peace, and Mayor of Tin Cup.

Dan and Julia Harrington had a large zinc bathtub — the only bathtub in town. Mrs. Bley supported her two children by washing and ironing the clothes of the dancehall girls — ruffle after ruffle after ruffle. Doc McGowan fit the mold of a frontier doctor. He carved his own crutches, drank heavily, and nearly starved for lack of business. Dry goods and supplies were purchased from such merchants as Carl Freeman, Sam Gollagher, John Weston, Will Woll and Will Niederhut. William R. Kreutzer became the first Forest Ranger in the United States. His residence, which was also his headquarters, is presently the Tin Cup Store on Grande Avenue. Kreutzer Peak is named in his honor.

Tin Cup is situated at an elevation of 10,182 feet. Each year, part of September was spent gathering firewood and canning goods for the long hard winter. Temperatures sometimes dropped to 40 or 45 degrees below zero. Stagecoaches, wagons, and carriages came to a halt through much of the winter.

In 1905 it was announced that a horseless carriage — an automobile — was coming to Tin Cup. This was quite an occasion, considering the remote and isolated location of the community. The whole town turned out to see the automobile "drive" in from Pitkin. Well, it didn't arrive, and still didn't arrive, but people continued to wait. Finally, everyone heard the engine, and amid much celebration the automobile chugged into Tin Cup. There was an explanation for its tardiness. The automobile couldn't negotiate the high altitude of Cumberland Pass, so a team of horses had to pull it over the pass to within two miles of Tin Cup where the horses were unhitched and the motor cranked. The horseless carriage was only horseless for a short portion of the trip.

Outside Gollagher's Store stood a storage shed from which kerosene was sold. In 1906 a fire started in the shed, then swept down Washington Avenue destroying everything for one city block before it could be brought under control. Another fire occurred across the street on the north side of Washington Avenue in 1913 which heavily damaged several buildings. Lost were the LaTourette Saloon building (where the fire started), the Tin Cup Mercantile Store, Frenchy's Place (his saloon and residence) and the Frank Beyers building.

Art Napoleon "Frenchy" Perault had living quarters in the back of his saloon. Frenchy was in bed sick with pneumonia when the 1913 fire broke out. When friends came to the rescue they tore the back door from its hinges. Frenchy was short of stature but very, very overweight. Not only was he too sick to move, but he was heavy. Rescuers rolled him onto the door and dragged him to safety. Having been wiped out, Frenchy Perault, who established Tin Cup's first saloon in 1879, retired. He died two years later in 1915.

Tin Cup has a unique cemetery. There are four sections — each on a separate knoll. Protestants are buried on the largest knoll. There is also one for Catholics and another for Jews. The fourth and highest knoll is called Boot Hill, the resting place for men gunned-down during Tin Cup's wilder days. One dance hall girl may also be buried at Boot Hill. Winter burials usually took place in Buena Vista, for the ground at Tin Cup was too frozen for digging. Funerals were allowed to travel toll free over Cottonwood Pass. Mourners paid toll on their return.

Captain Hall and Carl Hord didn't keep the Gold Cup Mine for long. The Bald Mountain Mining Company bought the property for nearly $400,000. The Gold Cup became the top producer in the Tin Cup District ($7,000,000). So rich was the ore that loaded jack trains on their way to the smelter were protected by several heavily armed guards. The Little Gold Cup, the Silver Cup, and the Addie were other good producers of the Bald Mountain Mining Company. Another successful mining company was the E. C. Stoddard Company who owned and operated the rich Jimmy Mack Mine. Other mines in the vicinity that did well were the Blistered Horn, Iron Bonnet, Drew, Tin Cup, Mayflower, Anna Dedricka,

El Capitan, and later the Bonnie Belle. Although there has been some activity since, all the working mines had shut down by 1918.

Tin Cup is located at the south edge of Taylor Park. It looks out at some of the most beautiful countryside in Colorado — pine-covered hills and snowclad peaks. The Town Hall and many cabins remain — and look much as they once did. Tin Cup today is a haven for summer residents.

Initially, Pitkin had 180 dogs but only 3 women — a crisis which didn't last long, however, as girls flocked in to "work" the saloons and gambling dens. Respectful women arrived as well.

PITKIN

Location: 27 miles east of Gunnison via U. S. Highway 50 and State Highway 162

Two fortune seekers had been prospecting near the future site of Pitkin for nearly two years without luck. One was striking a rock with his hammer as they were discussing plans to pack up and move on. Suddenly a fragment broke off but it wouldn't fall to the ground. The prospectors investigated this phenomenon and saw that wire silver was holding the chunk to the larger rock. They staked a claim — and canceled their plans to leave. The ore assayed at 80% solid silver. The Fairview Mine produced for 15 years.

More early strikes were made, including the Red Jacket, Iron Cap, Silent Friend, Silver Islet, and the Silver Age. The camp, initially called Quartzville, was established in April 1879. The first boardinghouse was a large tent with sawdust covering its dirt floor. There were rows of double bunks, each filled with hay, and the charge was fifty cents per bunk per night. A merchant sold $3,000 worth of hardware from his wagon while his store was being constructed. Saloons even operated on Sundays.

By the fall of 1879 many permanent buildings had been erected, and a substantial town was under way. The town's name was changed to Pitkin in honor of Governor Frederick W. Pitkin.

By 1881 the community had a population of 1,500. There was a bank, several hotels, cafes, and various types of shops. A newspaper was established, the *Pitkin Independent*. A second paper, the *Pitkin Mining News*, came later. The town also had many saloons, gambling dens, and brothels. It was a lively and rowdy place — especially during construction of the Alpine Tunnel. Railroad workers spent many nights "whooping" it up in Pitkin. On one occasion, in a dispute arising from a card game, the dealer was shot and killed by a railroad worker. According to one newspaper, the killer was "jerked to Jesus," at the first legal hanging in Gunnison County.

In addition to the aforementioned mines, many others were located which helped bolster the economy. The Tycoon, Little Roy, Little Addie Addie, Western Hemisphere, Nest Egg, and the Wampum were some of the more successful mines. The greatest boost to Pitkin occurred in 1882 when the Alpine Tunnel opened. Trains of the Denver, Southpark & Pacific Railroad rolled in from Hancock, Saint Elmo, and points east. Eventually the tracks continued on to Parlin and Gunnison.

By 1891 Pitkin had two churches and more women, two elements which brought a greater degree of respectability and culture to the community. The Colorado State Fish Hatchery had been established below town. A new rich strike at the Little Tycoon, coupled with heavy production from the Silver Islet and others, helped sustain area mining. The silver crash of 1893 dealt Pitkin a crippling blow, but many determined miners kept plugging along and weathered the storm. In fact, after the crash, more money was invested into the Fairview, Tycoon, Nest Egg, and Little Roy. A new city hall was constructed in 1901, and Pitkin received telephone service in 1904. Damaging fires in 1898 and 1903 destroyed much property. Through all the ups and downs, Pitkin persevered and today it is a quiet town in a lovely setting.

OHIO CITY

Location: 20 miles east of Gunnison via U.S. Highway 50 and State Highway 162

Some placer gold was located along Gold Creek (originally called Ohio Creek) and in the adjacent German Flats area during the 1860s. Deposits seemed shallow however, and the prospectors drifted away. Nothing much happened until rich deposits of silver were discovered in 1879 — and the rush was on. At the junction of what is now Gold Creek and Quartz Creek a camp was established which was briefly known as Eagle City, then changed to Ohio City.

A large hotel (and restaurant) was constructed, as well as several stores and saloons. A stageline connected Ohio City with Pitkin. In 1882 the Denver, South Park & Pacific Railroad dropped down from the Alpine Tunnel through Pitkin and arrived into Ohio City. It eventually continued on to Parlin and Gunnison. The community profited from the arrival of the D,S.P.& P. The silver crash of 1893 was a blow to Ohio City. Miners packed up and moved on. The community rebounded quickly, however, when gold was once again discovered in 1896. Another boom occurred and more new buildings went up. It is interesting that the Ohio City of the '80s was constructed entirely of logs, but by the time the second boom occurred a sawmill had been established at Pitkin and all the new construction was of finished lumber and frame.

The Eagle, Calumet, and Roller loads were adjacent to the city. The most productive mines however, were along Gold Creek. The Carter Group of mines shipped a gold brick worth $3,000 every other week. Further up the gulch lies the remains of the Raymond Mine, which yielded over $7,000,000 by 1916. Beyond were the properties of the Gold Link Mining Company which included 6,000 acres of claims, a forty-stamp mill, a 4,000 foot tunnel, a two-story boardinghouse, and many other buildings. They maintained a payroll of about 200 men.

East of Ohio City, on the road to Pitkin, lies a gravesite. The epitaph once scrawled on the "headboard" read:

> "Death went prospecting
> And he was no fool
> Here he struck faithful Pete
> The emigrant mule."

In the 1880s a gun battle occurred in which both participants had very poor aim. Two fellows named Reid and Edwards were separated after a heated argument in Leadville. Later they spied each other nearly a block away on an Ohio City street. Both men drew, and as they approached each other fired many shots. Finally, the bullets hit simultaneously into each others' heart. They were so close when they dropped that they almost touched each other.

The _White Pine Cone_ of George S. Irwin once stated, "White Pine suffered an agonizing famine this week. For two whole days there was not a drop of whiskey in town. Nothing but a liberal supply of peach brandy and bottled beer prevented a panic."

WHITE PINE

Location: 43 miles east of Gunnison via U. S. Highway 50 and Forest Route 888

Rumors of gold, created by the "Legend of Snow Blind Gulch," kept picks and shovels active for several years prior to the first recorded strike in 1878. According to the story, two prospectors discovered gold so rich that they were each washing out a pound per day. Their greed bested their common sense, and they were trapped by a winter snow storm and ultimately became snow-blind. The story may have been true for in 1878 prospectors found the remains of sluices and an old whipsaw along Tomichi Creek in Snow Blind Gulch. Many strikes, mostly rich in silver were made during the first year. A trio of mining camps — White Pine, Lake's Camp (North Star), and Argenta (Tomichi) — cropped up in the vicinity. The history and development of each paralleled the others. White Pine became the largest of the three camps.

White Pine (often spelled as one word, Whitepine) boasted several productive mines. The best were the May-Mazeppa, Morning Star, Evening Star, Black Warrior, Akron, Erie, Copper Bottom, and the Copper Queen. Top producer was the North Star Mine located to the east (see North Star).

In 1885, White Pine had neither a church or a preacher, so Sunday services were a little unusual. Professor Turner "preached" at the schoolhouse each Sunday. His talks were pleasant, and nothing was said to offend either saint or sinner.

Although another newspaper preceded it, the *White Pine Cone* began in 1883 and was in circulation for ten years. The spice and humor of editor George S. Irwin made the newspaper a hit throughout the area. On December 20, 1892, Irwin made the following plea: "Something over $1,000 is due the *Cone*. We need half of it. Please pay what you can on account." That was the last issue to roll off the press.

At first the business district of White Pine was comprised of the post office (established August 12, 1880), a two story hotel (constructed of logs), a blacksmith shop, and the usual shops and saloons. Later the Grand Hotel, the City Hall, and the Tivoli Theatre were constructed.

Mining, which was in decline after 1890, was dealt a near fatal blow by the devaluation of silver in 1893. White Pine rebounded in 1901, again in 1941, and once again in 1947. In fact someone will probably always be swinging a pick somewhere near White Pine.

NORTH STAR

Location: 1 mile northeast of White Pine

Just over one mile above White Pine, high above Galina Gulch toward the Continental Divide, is the site of North Star. It was originally called Lake's Camp, and later renamed for the North Star Mine which was located in 1879.

The camp was actually built on property of the May-Mazeppa Mine. Its fortunes, however, seemed to follow those of the rich North Star Mine. The devaluation of silver in 1893 temporarily put the North Star "out of business." The mine reopened a few years later, however, and the camp experienced a new boom. During 1901 three eight-hour shifts worked the mine daily. More new buildings popped up including a hotel, the Leadville House, whose false-front faced Main Street. Though mining declined once more, there was some activity well into the twentieth century.

The Soup Bone Musical Club was the pride of North Star. The group provided festive song on many occasions. And they did so with the accompaniment of a bell harp and guitar.

TOMICHI

Location: 2 miles north of White Pine

The Magna Charta Tunnel was the best of several good producing silver mines. The underground drifts and cross cuts stretched for a mile and a half and nearly to the communities of North Star and White Pine. There were other fine producers — the Eureka, Little Carrie, Lewiston, Brittle Silver, and Sleeping Pet.

Adjacent to the Magna Charta Tunnel, a settlement sprang up. It was originally called Argenta, later Tomichi Camp, and then shortened to Tomichi. By 1880, the community had a population of 1,500 and boasted that it was larger than White Pine.

Life was tough in most mining camps and bad luck commonplace — Tomichi was no exception. A fine smelter was constructed to serve the area mines. Its life was short, however, as it was destroyed by fire in 1883. Once, when an order of paper stock failed to reach the *Tomichi Herald*, the newspaper was printed on wrapping paper. A mining company had just tunneled to a rich silver vein when the silver crash of 1893 struck. Panic hit, and miners evacuated the camp. All the mines closed with the exception of the Eureka which operated until 1895. A few stragglers returned to rework some of the properties. In 1899 a snow slide killed five or six people and wiped out the town.

Nothing is left of Tomichi except for a few traces of the Magna Charta on the side of Granite Mountain.

ALPINE STATION

Location: 8 miles northeast of Pitkin at the west end of the Alpine Tunnel.

Alpine Station began to grow as a railroad construction town in 1880, during the year in which boring began on the Alpine Tunnel project. A stone engine house, turntable, and telegraph shack were built in 1881. Although the tunnel was completed by the end of 1881, construction continued for another four years. A snowshed 650 feet long was erected at the western end of the tunnel. By contrast, Atlantic, at the eastern end, required a snowshed of only 150 feet. A stone section house, a bunkhouse, and a storehouse were also built at Alpine Station.

The 11,546 foot elevation and the rugged winters in the Sawatch Range were not conducive to year-round living. After construction was completed, most of the workers had gone. Alpine Station remained as a railroad stop on the Denver, South Park and Pacific Railroad.

Avalanches and stranded trains were commonplace. On one occasion, in

March 1884, an avalanche caused by a train whistle swept away the little town of Woodstock which was located in a high meadow below Alpine Station. Fourteen of the seventeen inhabitants lost their lives.

Part of Alpine Station had to be rebuilt after a fire in 1906. Alpine Station was short-lived however. The huge and expensive beams of California redwood, which were used to construct the tunnel, failed. In 1910 there was a massive cave-in claiming several lives. The tunnel was never reopened, and Alpine Station became a ghost. A sign on the remains of the stationhouse reads, "Built in 1884 by DSPP RR."

HILLERTON

Location: 2 miles north of Tin Cup via Forest Route 765

North of Virginia City (Tin Cup) lay two "suburbs" — Hillerton and Abbeyville. Both were short-lived. Abbeyville was a small camp, midway between Hillerton and Virginia City, which cropped up, around the C. F. Abbey Smelter in 1881. Hillerton had a fast start in 1879, quick life, and then died rapidly. The community was named for one of it's founders, Edward Hiller.

Several strikes were made — including the Adeline, Little Earl, What Is It, and more — and a new toll road was opened to the stage stop at Jack's Cabin, the shortest route to Crested Butte and Irwin. Many businesses sprang up including a hotel, The New England House, and Edward Hiller's bank. Hillerton offered promise for the future, and hundreds of miners flocked in.

Henry Olney started a newspaper, the *Hillerton Occident*, during June 1879. Realizing quickly that Virginia City offered more promise than Hillerton, he packed up his press four short months later and moved it two miles south to start a new newspaper there. Olney wasn't the only one who realized "the grass was greener" two miles south. The mines at Hillerton were underdeveloped, while those near Virginia City were large and producing well. Residents carted their cabin logs south to Virginia City where they rebuilt their homes. Nothing remains today of either Hillerton or Abbeybille.

An architectural design unique only to Crested Butte is the two-story outhouse. The upper story is offset from the lower so persons on both floors could use it simultaneously.

CRESTED BUTTE

Location: 28 miles north of Gunnison on State Highway 135

Crested Butte never experienced the boom and bust pattern typical of so many mining towns. Probably because of its continuous diversification, its growth was steady.

The town actually started as a gold camp when nuggets were first discovered in nearby Washington Gulch in the '70s. It developed as a supply center, then really emerged as a coal mining community. At the time it was the only location west of Pennsylvania where bituminous coal was mined.

By July 3, 1880, the incorporation date of Crested Butte, there were over fifty buildings, a newspaper — the *Crested Butte Republican*, and 250 residents. Also, the narrow-gauge railway was on its way from Gunnison.

With the completion of the railroad and the opening of the road over Pearl Pass to Ashcroft and Aspen, Crested Butte became an important travel center. Travelers frequented the new Elk Mountain Hotel (completed in 1882) which boasted of its fine cuisine.

The town was given an extra boost when the Colorado Fuel and Iron Company assumed operation of some of the mines in 1882 and 1883. There were three bituminous mines, three anthracite mines, and 150 coke ovens. Coal sustained the community for nearly seventy years until the last mine, Big Mine, shut down in 1952. The economy languished for over a decade before Crested Butte developed as the fine ski resort and tourist center that it is today.

In 1974 the entire town was designated a National Historic District. As the community grows in the 1990s it is doing a remarkable job of blending new and dated Victorian architecture to keep the look "a century old."

IRWIN

Location: 8 miles west of Crested Butte. Take the right fork at the Lake Irwin Loop.

Irwin was one of those mining towns that came fast, went fast, and, during the interim of five years, lived fast.

Dick Irwin made a rich strike during the late fall of 1879. As word got out the ensuing winter, fortune-seekers flocked to the vicinity. Even with heavy snow on the ground they cut trees and built cabins. When the snow melted in the spring there were ten-foot tall tree stumps everywhere.

Initially a few small camps were built which eventually became absorbed by Irwin. One of these "suburbs" was platted by a con-artist from Leadville. He promised that he would build a six story hotel, a theatre, an office building and a grocery store in the center of Ruby City. After selling many lots, he pocketed the money and vanished.

In a short two years Irwin had twenty-three saloons, several hotels, the usual parlor houses, gambling halls, and even fire hydrants. Main Street was a mile long.

Legend has it that a plot was discovered to assassinate former President Ulysses S. Grant during a forthcoming speech in Irwin. The speech was canceled and Grant was entertained for two days at the ornate and exclusive Irwin Club

— a men's club of Irwin's elite. Other notable guests of Irwin were Teddy Roosevelt and Wild Bill Hickok.

The owner of the Forest Queen, the closest mine to the townsite and one of the richest, once turned down an offer of $1,000,000 for the mine. In 1932 it sold for taxes. The buyer paid $40.45.

Although a little mining activity occurred later, the town virtually died in 1884. People left so quickly, it is told, that some of the cabins still had dishes on the tables.

A few old splinters and a few new cabins mark the location of Irwin. Even the fire hydrants have disappeared.

In an area where it was easier to dig graves, the Irwin cemetery stands two miles below the townsite where the road to Keebler Pass climbs the hill immediately above the intersection of the road to Ohio Pass.

"Best way to find gold in the summer is to do your sleeping in the winter."

GOTHIC
Location: 8 miles north of Crested Butte via Gothic Road

Below the pyramided cones of Gothic Mountain (for which the town was named) both gold and silver ore were discovered in June of 1879. Prospectors swarmed to the area by the hundreds. Within four months, 170 buildings had been constructed including two sawmills, a hotel, three stores, a butcher shop, and a saloon.

Gothic became the richest city in Gunnison County. Millions of dollars worth of gold and silver were taken from the mountains — some of it valued up to $15,000 per ton.

Along with the riches, Gothic gained the reputation of being one of the wildest cities in Colorado. The image was due in part to the scantily-clad dance hall girls who wore skirts all the way up to their knees.

In 1880 Ulysses S. Grant wanted to see some of the boom towns. He headed to Gothic which was in its glory at the time. He drove his own stage into town to a noisy, gun-popping celebration.

The city had a most unusual way of "electing" its first mayor. Lew Wait, a newspaperman, and Garwood H. Judd, a saloonkeeper, were both interested in the job. They rolled dice — and Wait, who edited the *Gothic Silver Record*, won the job.

Judd later gained fame as the "man who stayed." He remained in Gothic years after everyone else had left. Fox Films made a two-reel picture about him in 1928.

The population swelled to about 8,000 at its peak. By then there were three sawmills, several hotels, a smelter, two school houses, and many stores. The city became a supply depot for smaller camps in the vicinity, such as Schofield and Elko.

Weekends were a time for fun. Horse races were held every Sunday on the edge of town. Dances were a regular occurrence with music furnished by Gothic's own four piece orchestra.

The heavy snows and steep mountain slopes presented many problems. Many people lost their lives in snowslides, which were commonplace. Families were often snowed-in and wagons marooned — but it was a way of life. The townspeople wore snowshoes much of each year.

By 1884 most of the gold and silver had disappeared — as well as the people. The setting is beautiful and several old buildings remain. It is a town worth seeing. Gothic has been revived as a summer residential area.

SCHOFIELD
Location: 3 miles southeast of Crystal

In 1872 a group of prospectors found silver bearing ore near the head of Rock Creek. During the following spring they established a small smelting furnace at the location to test their ores. Excited about the prospect of what they had found, they traveled to Denver to raise mining capital. Their plans were sidetracked by the uncertainty of the silver market in the fall of 1873. They never returned to Rock Creek.

In 1879 the mines near Rock Creek were discovered once more. A trio of camps sprang up along the Crystal River — Schofield, Elko, and Galena. Schofield was located in Crystal Canyon on the edge of a high mountain meadow. Elko was south of Schofield and nearer the base of Galena Mountain. Galena was located about midway between the two. Neither Elko or Galena amounted to much. All three were very inaccessible. Winter lasted eight months of the year, with snow sometimes forty feet deep. Initially a terrible road was blazed from the south up and across Schofield Pass and down to the camps. At the time it was the only way in.

Schofield was platted in August 1879 by B. F. Schofield and his group. A smelter was built in 1880, and a mill in 1881. By the following year the community had a hotel (which was operated by a woman from Illinois), restaurant, a general merchandise store, blacksmith shop, and even a barbershop. When the road was completed up Crystal Canyon from Crystal to the northwest, Schofield became a stagecoach stop.

During the summer of 1880 Schofield had a colorful visitor — former President Ulysses S. Grant. Grant was accompanied by his son and former Governor Routt. There wasn't much to see in Schofield in 1880, but residents rolled out a barrel of whiskey and Grant had an enjoyable visit.

Old Lady Jack was one of Schofield's most unusual inhabitants. She wore a gunnysack as a shawl, and claimed that famed Indian scout Jim Bridger was her uncle. Old Lady Jack had a fetish for cleanliness and cats. She washed her mob of cats often, then hanged them to dry by the napes of their necks. It is said, that she even washed her firewood in order to keep her hands clean.

The mountains yielded average ores, but the time and cost of transporting them was a major problem. So were the long harsh winters. The life of Schofield was short — by 1885 most everybody had left.

CRYSTAL
Location: 5 miles east of Marble

Crystal is located in a spectacular setting. The townsite lies adjacent to the Crystal River in a high mountain meadow in Crystal Canyon. It is surrounded by towering peaks — Mineral Point, Treasury Peak, Sheep Mountain, Bear Mountain, and Crystal Mountain. On the opposite bank of the Crystal River stands the picturesque remains of the Crystal Mill (built by G.C. Eaton) which once provided power for the Sheep Mountain Tunnel.

Crystal was a silver town which began in 1880 when some abundant strikes were made. The first trail to the area came from Crested Butte (17 miles away) by way of Schofield Pass. Because of this poor access, only the richest ores were packed out during the early development of the mines. Transportation improved when the road to Marble extended on to Carbondale. The Lead King Mine boosted the economy of both Crystal and Marble. In fact it bounced back after the silver crash of 1893 and continued to produce until 1913. Silver from the Black Queen Mine was exhibited at the Chicago World's Fair in 1893. Other productive mines were the aforementioned Sheep Mountain, the Lead Queen, Black Eagle, Inez, and the Catalba. The mines not only produced silver, but lead, zinc, and a small amount of gold.

Al Johnson was one of the citizens most instrumental in the development of Crystal City — the original name of the camp. In 1880 he began publishing the *Crystal River Current*, and hired newspaperman Tom O'Brien to manage it. Johnson owned the hotel, the store, and a jack train which packed ore to Crested Butte and brought back supplies. He also operated the post office while his brother Fred carried the mail. Crystal was often snowbound during the winter, and Fred Johnson carried the mail to and from Crested Butte on snowshoes. As Crystal continued to grow, a second hotel was constructed, as was a schoolhouse. There were several stores and saloons — including the exclusively male Crystal Club (which still stands). The population which peaked at about 500 dwindled to only a handful after the panic of '93.

In 1954 Mr. and Mrs. Joseph Neal and Mrs. Helen Collins, started a program to restore and preserve the remaining buildings. A few summer residents, and their jeeps, keep Crystal alive.

"According to legend there is a vein of ore worth $5,000 per ton somewhere in a lost cliff near Muddy Creek."

MARBLE

Location: 11 miles south of Redstone via State Highway 133

Yule Creek, the settlement that was to become Marble, became a reality because of mining, not marble. The success and economy of the community, however, were a result of marble, not mining.

Actually there were two camps, Yule Creek and Clarence. Both blossomed in the early 1880s, then grew together to form Marble City — then Marble. The Lead King Mine was the area's best producer. It operated during the '80s and '90s yielding silver, lead, and some gold. A smelter was constructed in 1890 by the Hoffmans. Coke for the furnace was hauled all the way from Crested Butte, over Schofield Pass, down through Crystal to the smelter. A newer Hoffman smelter was built in 1901.

The massive marble deposits on White House Mountain had been general knowledge for years, but logistics precluded their development. Several quarries opened during the 1890s — one by J.C. Osgood of Redstone fame — but it was not until after the turn of the century that the marble industry shifted into high-gear. Nearly pure white marble was shipped near and far. The stone was furnished for some very notable structures such as the Lincoln Memorial in Washington D.C.; the Tomb of the Unknown Soldier in Arlington, Virginia; government buildings in New York, San Francisco, and Cleveland; as well as several buildings in Denver — including the Colorado National Bank, the Colorado State Museum, the Customs House, and the Cheesman Park Pavilion. It took 75 workers at the Colorado Yule Marble Company quarry one year to cut the enormous stone required for the Tomb of the Unknown Soldier. The block weighed 56 tons after it was trimmed.

Sylvia Smith, who operated the *Marble City Times,* took delight in attacking the Colorado Yule Marble Company. She blamed the company for fraud and stock manipulation. Most of her complaints fell on deaf ears however as the company was Marble's largest employer with a work force of up to 1,000 workers. The last straw was her "I told you so" arrogance after a mine disaster. She, and her printing press, were run out of town.

As the market increased for marble substitutes and low-cost marble veneers, the quarries fell on hard times — so did the community of Marble. Adjacent to the Crystal River stand the marble columns of the once productive mill — looking much like pillars that should be standing in Athens or Rome.

CHANCE

Location: 9 miles southeast of Gunnison

Mineral Hill, where most of the mining properties were located, separated Chance and it's neighbor Iris. Iris, just over one mile away across the Sagauche County line, was the larger of the two camps, but Chance produced more gold. Gold production was below par and unprofitable however, and Chance lived a short life. The camp was established in 1894.

In four years all but one of the mines had stopped production. The best mines were the Lucky Strike and Only Chance.

The combined population of the two camps was about 1,000. Chance had, as did Iris, several stores, saloons, a blacksmith shop, livery, and hotel. Chance had tri-weekly mail delivery and telephone service from Gunnison.

A great deal of capital was invested into area mining. Much of it was to little avail, however and Chance fell on hard times. The mill of A. E. Reynolds was constructed at a cost of $50,000 but lasted just one year before it closed. Total production through the mill never exceeded $30,000.

VULCAN

Location: 26 miles south of Gunnison

Following the discovery in 1895 of the Vulcan, Mammoth Chimney, and St. Patrick mines, a town was platted, and Camp Creek sprang into being. Several hundred prospectors flocked into the vicinity. The community was renamed Vulcan in 1897 for nearby Vulcan Hill, an extinct volcano.

A fire once raged in the shafts of the Vulcan Mine for several days before it could be brought under control. The mine, which was a top producer, eventually consolidated with the Good Hope and Lincoln. Jointly they produced a half-million dollars in gold. Later the Mammoth Chimney joined the group which was then operated as the Vulcan Mines and Smelter Company.

Through its flourishing years, population of the settlement remained at about 500, of which nearly one-third worked at the Vulcan Mine. It was a hard working community in which everyone looked forward to Saturday night dances with great anticipation. There were several saloons, shops, a hotel, a smelter, and one newspaper — the *Vulcan Enterprise*. The *Vulcan Times* was published after the turn of the century but was short lived. School classes were conducted in an old store five months of the year.

The community was besieged by a flurry of union troubles in 1899 but persevered. Many different minerals were shipped from Vulcan during its heyday. As was the case with many mining towns, the deposits played out — and the town faded.

BALDWIN

Location: 17 miles northwest of Gunnison via State Highway 135 and the Ohio Creek Road

Would anybody like to buy a ghost town? Baldwin, with it's sagging buildings, creaking floors, and banging shutters, is fenced with a locked gate, and has a sign which reads, "For Sale — 1518 Acres."

Although its post office was established in 1883, the ghost-town of Baldwin which exists today didn't spring into being until 1897 — when gold was discovered nearby. Evidently the Baldwin post office originated at a little community to the south (later to be known as Castleton). When the newer town site was established, it was known briefly as Citizen and then Mt. Carbon. As it began to flourish, "old" Baldwin was dying. Following a shift in coal mining operations, on June 26, 1909, the post office at Mt. Carbon simply changed its name — and the "new" Baldwin was officially established.

The present site was located because of the gold strike, but its rewards were a result of coal mining. Baldwin was a prosperous coal camp, but its history was marked with labor strife and casualties. Disputes resulted in many labor strikes. In separate incidents, a "scab" was killed by a union worker, a bridge was blown up by strikers, and a mine superintendent was killed. Mining continued, however, until World War II.

The post office was discontinued in 1949, as everyone had left — everyone that is except Joseph "Peanuts" Berta. Baldwin had been his home for years and he had no desire to move. Adjacent to the townsite is his tombstone, which was set upon his death in 1967.

"They never worked so hard in their lives to get rich without working."

DORCHESTER

Location: 11 miles north of Taylor Park Reservoir via Forest Route 742

Dorchester had a late-blooming and struggling existence. The camp sprang to life after several gold strikes were made in the Italian Mountains in 1900.

It was often difficult to work the mines because of the heavy snows. Occasionally it took weeks for a four-horse team to make a round trip to Aspen. Some of the mines worked through the winter which was quite hazardous due to the frequency of snow slides.

Regardless of the drawbacks, over a thousand miners and prospectors swarmed into the area for a while. Many left seeking an easier place to find their fortune. Dorchester was also a stopover for those traveling between Tin Cup and Aspen, but it never really amounted to much as a town. Some mining activity continued through World War I, but after the mines were closed the camp was deserted.

"If you have to 'prove' you're right, you're probably wrong."

BOWERMAN

Location: 3 miles north of Waunita Hot Springs

J. C. Bowerman was a prospector down on his luck. His wife did odd jobs to keep food on the table and provide him with supplies. It wasn't enough. Bowerman, out of desperation, advertised for a grubstake.

A railroad man responded to his ad, offering Bowerman $50 per month for supplies. At this point a strange story began to develop. Bowerman struck rich ore said to be assayed at $70,000 a ton. Word got out, and suddenly there were hundreds of prospectors combing the hills. The town of Bowerman was established in 1903. By the following year there were two hotels, a newspaper, *The Bowerman Herald*, five saloons, five gambling halls, and several businesses.

Several other claims were established, most of which shipped low grade ore. Bowerman's mine, the Independent, had not shipped any ore out, however — and, this was supposed to be the rich find that brought throngs of people to the vicinity. Bowerman and his partner fenced the property, and wouldn't show the mine to anyone. They had all kinds of excuses for their failure to produce.

The five-hundred, or so, townspeople still had high expectations — for awhile. The other mines were "average", or below, as was the Independent once it finally started shipping. As people realized that the area was far below their original anticipation, they began drifting away. By late 1910 Bowerman was a ghost-town.

Tin Cup in 1906 shortly after completion of the town hall. *Collection of Dave Southworth.*

The saloon of Billy Reese at Pitkin. *Denver Public Library, Western History Department.*

The hotel at White Pine. *Denver Public Library, Western History Department.*

The *Pilot* newspaper office at Crested Butte in 1883. *Colorado Historical Society.*

GUNNISON COUNTY 269

The Crystal Mill sits on the river's edge amid spectacular scenery. *Dave Southworth.*

This general store and post office at Dorchester in 1909.
Colorado Historical Society.

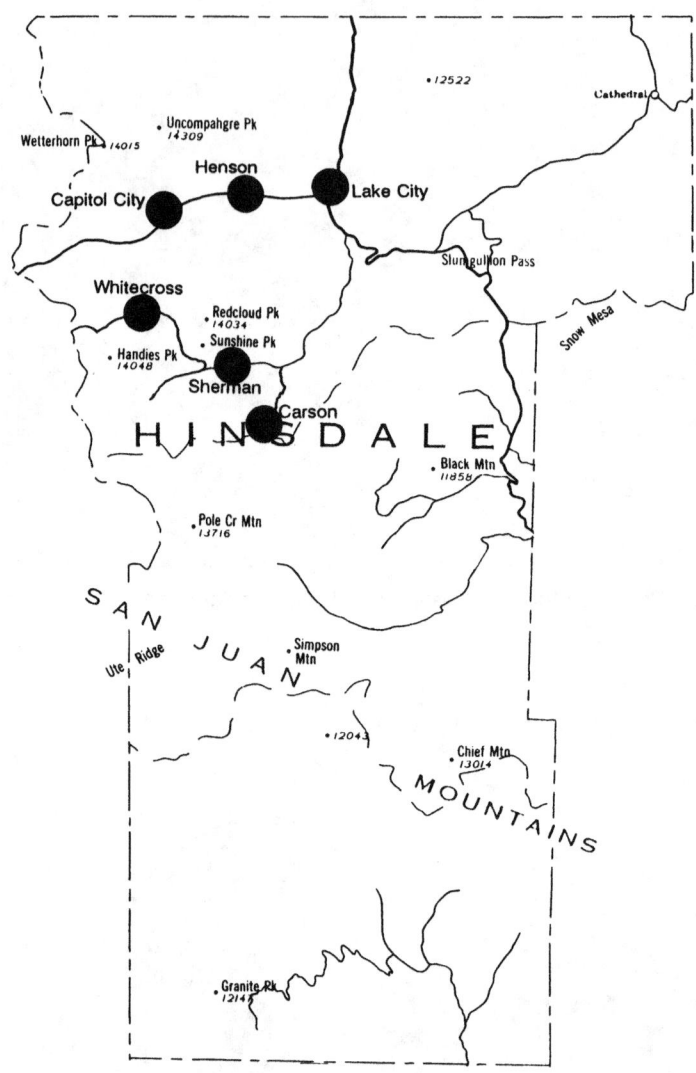

HINSDALE COUNTY

Around Lake City, it is told, there was a prospector's burro who actually wore snow shoes on treks into the wintery mountains.

LAKE CITY

Location: 55 miles south of Gunnison on State Highway 149

The Ute Indians occupied the area which was to become Lake City until the Brunot Agreement (which was ratified in 1874) opened the area for white settlers. The Ute-Ulay Mine, located in 1871 (see Henson), was the first gold strike in the area. The mine, located west of town, could not be developed until after the relocation of the Indians. The Golden Fleece (originally named the Hotchkiss Mine for its discoverer Enos Hotchkiss) was discovered in 1874. The mine, which had many ownership changes, was the best producer in the area. The Belle of the West and other strikes followed. The development of Lake City was well on its way.

By 1876 the community had four hotels, seven saloons, a newspaper, the *Silver World*, and the first church on the western slope — the Presbyterian Church. A reduction plant, which was powered by a seventy foot waterfall, was constructed above the community at Granite Falls.

Lake City had a wild red-light district. In Hell's Acre, as it was called, gambling dens and dance halls were mixed with the many brothels and the Crystal Palace — the posh bordello of Clara Ogden. One of the joints was owned by two characters named Betts and Browning, who had a propensity for thievery. In 1882, during a confrontation with the law, the pair shot and killed Sheriff E. N. Campbell. Betts and Browning fled but were captured by a posse and jailed. The community was irate. A raging lynch-mob dragged the pair out of jail and hung them from a nearby bridge.

Lake City had a legendary preacher, Rev. George M. Darley, who documented his tribulations. He preached the gospel in the most unlikely places. Darley believed that the folks who needed religion the most were the ones who lived in, and frequented, Hell's Acre, so he spread the holy word in gambling dens and dance halls.

Another church was built. Two more newspapers began publishing, the *Mining Register* and the *Lake City Phonograph*. Two banks and a library were constructed, and the stone schoolhouse was built in 1882. People came from great distances to shop at Lake City's large variety of stores. The city also had one of the earliest telephone systems in the state.

After the Denver & Rio Grande arrived in 1889, mines near some of the high mountain camps such as Whitecross, Sherman, and Carson shipped their ores by wagon as far as twenty five miles to reach the railroad at Lake City.

Lake City, which had a population of about 4,000 at its peak, was a socially active community. Dances, concerts, and banquets were commonplace. There were several clubs, civic organizations, and many other group functions. Much of the social activity centered around the churches and the Occidental Hotel.

In 1874 at the foot of Slumgullion Pass, the remains of five men were discovered. One had been shot, the other four had their skulls crushed. Each of the bodies had been carved up and fleshy parts removed. Herein lies the legend of Alferd Packer — that part of history for which Lake City is most famous.

Packer and five men were prospecting in the San Juan Mountains during the winter of '73-74. During the spring, Packer arrived alone at the Los Pinos Indian Agency near Saguache, stating that the others had left him and that he nearly starved on his trek back to civilization. After the bodies were found, Packer was arrested and charged with cannibalism. He escaped but was arrested again nine years later in Wyoming and returned to Lake City for trial. He received the death sentence. He won a new trial in Gunnison and received a prison term, which several years later was followed by a pardon. *Denver Post* newswoman, Polly Pry, who had crusaded for his release, was instrumental in getting Packer a job as doorman at the Post.

No other incidents of cannibalism have ever been tried in the United States court system. When sentence was first pronounced, according to legend and poet Stella Pavich, the judge told Packer: "...There was siven Dimmycrats in Hinsdale County! But you, yah voracious, man-eatin son of a b____, Yah et five of them, therefore I sentence ye T' be hanged by the neck ontil y're dead, dead, dead!" The comical and often-repeated quotation was not a statement from Judge M.B. Gerry, an articulate gentleman.

A commemorative to the men lies at the foot of Slumgullion Pass. Isn't it strange that the site where Alferd Packer allegedly had his meals is named after a miner's stew-slumgullion?

One of Lake City's most famous visitors was Susan B. Anthony. In 1877 a huge crowd gathered around the courthouse steps to hear her lecture on women's suffrage.

Lake City, which was named for Lake San Cristobal, never became a ghost-town — probably because of its many stone structures. The population dwindled to a couple of hundred people. The railroad tracks were removed in 1937; the Golden Fleece Mine was sold for taxes in 1943; the Occidental Hotel burned down in 1944; but Lake City has emerged as the peaceful and lovely community that it is today.

During one of their trips to Washington to partake in negotiations, Chief Ouray and a party of Utes attended the zoo and saw elephants for the first time. They made up a word for them, which when translated meant "the big high animal with a tail at each end."

HENSON

Location: 7 miles west of Lake City on the road to Engineer Pass

Henry Henson, for whom the town is named, and his partners Joel K. Mullen, Albert Mead, and Charles Goodwin struck gold in 1871 and located the Ute-Ulay Mine. The discovery, which was on Indian land, brought more prospectors into the area much to the outcry of the Ute Indians. The white settlers temporarily vacated the area. In 1874 the Brunot Agreement was ratified, which opened the San Juan Mountains for settlement. Henson and his partners returned to their claims, were successful, and in 1876 sold to Crooke Bros. of Lake City for $125,000. Crooke Bros. added a lead smelter which operated on ore from the area mines. In 1880 the property sold again for $1,200,000. In 1882 a concentrator was erected. By 1893 the plant was the largest in the San Juans.

The town which grew up around the mines was very congested because the gulch was narrow, leaving little room for the many stores and houses that were constructed. In 1877 the Henson Creek Road was extended to the west, topping the mountain range near Engineer Peak, and connecting with the Animas Forks Road to Ouray and Silverton.

Henson was a tough town. During its short history, labor problems, accidents, and shootings were commonplace. Through a miscalculation in tunneling, shafts from the Ute-Ulay and Hidden Treasure mines made connection. An explosion of gas released in the tunnels resulted in thirty-six casualties. Eighty Italian members of the Western Federation of Miners Union struck at the Ute-Ulay and Hidden Treasure Mines in 1899. They were tough, armed, and physically drove off all others attempting to work. The Italian Consul and the militia were called in to coax the union workers into surrendering. This was accomplished. Three days after their arrest, the company made an announcement advising that they would employ no more Italians. All single Italian miners were ordered to leave the county within three days and all married men within sixty days. The Italians left peacefully. The community declined, however, and the post office was discontinued fourteen years later.

"A Self-made man often worships his creator"

CAPITOL CITY

Location: 10 miles west of Lake City on the road to Engineer Pass

Several good silver strikes in 1877 brought throngs of prospectors into the area. A townsite was platted, and tents were quickly replaced by substantial structures. George S. Lee built a sawmill to aid in the construction of Capitol

City. Saloons, hotels, restaurants, and a general merchandise store sprang up. A post office was established in May of 1877.

Lee also constructed a smelter in 1879 along Henson Creek to process ore from the many mines nearby. The Ocean Wave Mine, which was located near the smelter, was one of the better producers in the vicinity. Other top mines were the Morning Star, Capitol City, Yellow Medicine, and Polar Star. Much litigation over claim rights hampered the total production of the area.

George S. Lee built a large and elegant brick home. In the house, which even contained a ballroom and orchestra pit, he and Mrs. Lee lavishly entertained guests from near and far. It is said that Lee had a dream that his town would become the capitol of Colorado (hence the name Capitol City) and that his home would become the governor's mansion. History has shown that Mr. Lee was overly optimistic with regard to the growth of the community.

After telephone service was established, a unique musical recital was held in 1881. Residents of Capitol City, Lake City, and communities as far as Silverton, who had telephones could pick up their receiver at a specified time and listen to the concert. In those days party lines could be connected.

The devaluation of silver in 1893 had a crippling effect. A few residents remained, but most moved on. The community experienced a short revival when gold was discovered after the turn of the century. The post office was discontinued in October 1920.

On the road above Capitol City, at an altitude of 11,200 feet, is Rose's Cabin. Shortly after the Brunot Agreement (see Henson) opened up the area, Corydon Rose built a boardinghouse which housed a tavern and a restaurant as a stop for travelers on the road to Engineer Pass. Today, the remnant stands amidst a field of dandelions to mark the location.

SHERMAN

Location: 16 miles southwest of Lake City via State Highway 149 and the road to Cinnamon Pass (which follows the Lake Fork of the Gunnison River)

The site where the busy little mining camp of Sherman once existed is located in one of the most picturesque areas of Colorado. It is located in a lush wooded valley below Sunshine Peak — a fourteener. The scenic trail from Lake City to Animas Forks was originally a toll road. To or from Animas Forks required a fare of $2.00, while Lake City in the other direction was $2.50 either way.

Sherman began in 1877 because of successful placer and lode mining in the vicinity. The Black Wonder was the best mine, and produced well past the turn of the century. Other mines were the George Washington, Monster, Mountain View, Minnie Lee, Smile of Fortune, Clinton, and New Hope. The area yielded gold, silver, copper, and lead. Most of the ores were transported to the smelters at Lake City for reduction.

The townsite, which was platted with wide streets, was dominated by three buildings. There was the camp's only hotel — the Sherman House. A mini-mall of sorts housed a general merchandise store, grocery, butcher shop, bakery, and even a bunkhouse. The Black Wonder Mill was located in town as well. Sherman was a convenient stagecoach stop, for its location is nearly midway between Lake City and Animas Forks. Population of the community peaked at about 300.

Snare Creek, Cataract Creek, and other tributaries dumped into Cottonwood Creek which intersects with the Lake Fork of the Gunnison River at Sherman. The valley was continually troubled with floods as a result of the heavy spring runoff. In an effort to control the water, a large dam was constructed. Shortly after it was completed, torrential rains flooded the gulches — tore through the dam, and destroyed much of the townsite.

A Whitecross miner once stored blasting powder in an extra coffee pot. One morning while still sleepy-eyed he mistakenly placed the wrong pot on his stove.

WHITECROSS

Location: 21 miles southwest of Lake City via State Highway 149 and the road to Cinnamon Pass (along the Lake Fork of the Gunnison River)

Two veins of white quartz cross each other high on the face of Whitecross Mountain — hence the name of the mountain and the mining camp. A cluster of mining camps blossomed in the spectacular high alpine meadow of Burrows Park. Whitecross was the largest, and center of activity for the group which also included the camps of Tellurium, Sterling, Burrows Park, and Argentum. They all sprang to life between 1877 and 1880 following rich silver strikes in the vicinity. The post office was established in 1880 as Burrows Park (but was actually located at Whitecross). The name of the post office was changed to Whitecross in 1882. The community maintained a summer population of about 200. The Hotel de Clauson at Whitecross was the favorite meeting place for all of the neighboring camps.

The Tabasco, Champion, Cracker Jack, and Bonhomme were the top mines. The huge Tabasco Mill was constructed in 1901 just west of town. Burrows Park (the camp) had three good producers — the Undine, Napoleon, and the Oneida mines. The economy of Tellurium and Sterling relied on the Providence, Troy, Mountain King, Allen Dale and the Little Sarah. The latter group of mines either pinched-out early or were underdeveloped, and the short-lived camps of Tellurium and Sterling became abandoned. The railroad was just a few miles away at Animas Forks, but the road was so difficult that most ores were shipped to Lake City for reduction. Repeal of the Sherman Silver Purchase Act spelled doom for the silver industry and the remaining camps in Burrows Park became ghost towns.

CARSON

Location: Atop the Continental Divide, 11 miles south of Lake City

Carson straddles the top of the Continental Divide with parts of the camp on each side. Actually there are two sites — one higher and older which was a silver camp, the other lower and newer which was a gold camp. Situated at an altitude of nearly 12,000 feet, and covered with snow most of the year, Carson was often totally inaccessible.

Christopher J. Carson found traces of gold and silver in 1881 and staked his claim — the Bonanza King. Carson Camp (as it was first named) was established the following year.

Mines peppered the mountainsides. The St. Jacob's Mine was the area's top producer. Its yield in 1898 alone was $190,000. Other productive mines were the Thor, Maid of Carson, Legal Tender, Chandler, Kit Carson, and the Iron Mask, to name a few. Construction of the wagon road up Lost Trail Creek in 1887 facilitated shipment of ore out of, and supplies into, the area.

There are different theories as to the existence of two Carsons. Robert L. Brown's assessment seems most logical. Basically, he contends that the older Carson was predominantly a silver producing community — which it was. It probably faded after the silver panic of 1893. With new gold discoveries in 1896, the newer site sprang up near the Bachelor Mine, a gold producer. As the sites are close together, and whereas one was possibly abandoned when the other began, it is logical that they had the same name.

While the elements are taking their toll on the original Carson, the lower townsite is nicely withstanding the test of time.

The Presbyterian Church at Lake City — possibly the first church on the Western Slope. *Colorado Historical Society.*

The store and Black Wonder Mill at Sherman. *Collection of Dave Southworth.*

278 SOUTHWEST REGION

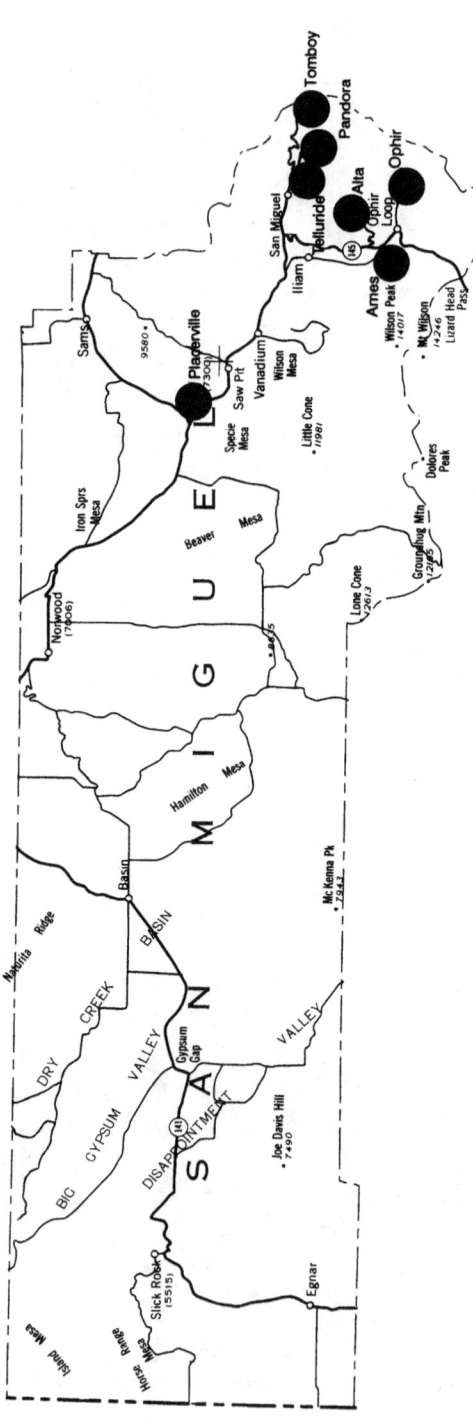

"It's harder to make a banker out of a thief than a thief out of a banker."

TELLURIDE
Location: 41 miles south of Ridgway via State Highways 62 and 145

Telluride is experiencing dramatic expansion as a year-round resort. The town itself, which is nestled on the floor of a gorgeous and spectacular box canyon, looks much as it did a century ago — with Victorian homes laced with gingerbread and topped with tin roofs. The scenic hills above, however, are in the midst of exciting development. Telluride has become a playground for many movie stars and other celebrities, and some have built beautiful homes in the area. As new resorts pop their heads above the treetops, the future looks very bright indeed.

It all began when John Fallon discovered the Sheridan vein ten miles east of Telluride, in August of 1875. The lode was tapped by other claims — the Union, Mendota, and Smuggler. The Smuggler was the result of a brilliant deduction by J. B. Ingram. Ingram thought that both the Sheridan and Union claims seemed too large. He discovered that both exceeded their legal allowance by several hundred feet. So, he staked the Smuggler on the area covered by the excess. The best producing mines during the '80s were the Smuggler, Mendota, Union, Argentine, Cleveland, Cimarron, Sheridan, Hidden Treasure, and Bullion. Following mergers and purchases, the Smuggler-Union emerged as the area's biggest producer. By the turn of the century, the company had 35 miles of tunnels. By that time, the Liberty Bell and the Tomboy had also emerged as top producers as well. Near the site of the Tomboy, the Japan and Columbia were located in 1894. In later years, several properties were purchased by the Idarado Mining Company. With advanced milling methods, Idarado became the largest producer of all.

During the late '70s, several small camps cropped up in the canyon along the San Miguel River. San Miguel City was the largest at first, but Columbia was closer to the mines and soon outdistanced all of the other camps. In 1881 the name was changed from Columbia to Telluride. Although its location was rather isolated, Telluride's growth was steady through the 1880s. The community, and the mining industry on which its economy was reliant, received a real boost with the arrival of the railroad in 1890.

Although Telluride was most noted for its lawlessness, it had another side also. To occupy the leisure time of those so inclined, there were many civic clubs, fraternal organizations, dances, and concerts. Ethnic groups had their own meeting places such as Swedish Hall and Finn Hall. Much advance preparation was made each year for the huge Fourth of July celebration. Fireworks, a huge parade, and a grand ball highlighted the festivities.

Whoever referred to Telluride with a play on words, "To hell you ride!", wasn't far wrong. According to historian Frank Hall, there were always "more than a sufficiency of saloons" in Telluride. Gambling dens such as Pacific Hall were quick to fleece a miner of his hard-earned wages — so were the brothels. Parlor houses were important in most all predominantly male mining towns, and Telluride was no exception. Its "sporting" establishments ranged from the Pick and Gad, a popular bordello, down to the simple cribs. A man with many enemies was the town marshal, Jim Clark. He was a crack shot who carried two pistols and stashed rifles at strategic locations around town in case he needed one in a hurry. In return for favors from storekeepers, Clark would bully debtors into paying their bills. While walking his beat one night, a shot rang out from the darkness between two stores and killed him instantly. Nobody knew who fired the shot — and, nobody really cared. In broad daylight one afternoon in 1889, with the help of two sidekicks, Butch Cassidy robbed his first bank in Telluride. The biggest bank heist, however, occurred years later and was an inside job. Following the crash of 1929 many banks failed. When Charles Waggoner, president of the Bank of Telluride, couldn't cover the deposits of his Telluride friends, he felt as though he had to do something quickly. By using certain banking codes, he had large drafts deposited by top New York banks to the credit of his bank. He covered the deposits of his Telluride friends, and in an effort to hide the rest of the money he scattered it across the country in smaller deposits. The fraud totalled a half million dollars. Waggoner was arrested and imprisoned, but to the people of Telluride he was a hero.

Much of the violence which occurred in Telluride was a result of the labor war. The Western Federation of Miners called a strike at the Smuggler-Union on May 2, 1901, to protest wages based on quantity of ore mined rather than the standard three dollar rate for eight hours which was common throughout most of the state. After six weeks of inactivity, mine owners hired non-union "scabs" at wages of three dollars per day, and reopened the mines. A confrontation occurred between 250 armed and irate union workers and the non-union men. After three men were killed and six more injured, the scabs were run out of town. The strike was settled and the union had won the first round. The mine owners won the second round, however, and they won it big. Tomboy miners struck in September of 1903 when the new mill was opened with non-union workers. Faced with a new threat of impending violence, Governor James Peabody declared a state of martial law. The militia seized control. Union laborers and union sympathizers were loaded in rail cars and run out of town. Many were beaten first. A statement was issued by the mine owners: "...We do not recognize a union in Telluride. There is no strike in Telluride. There is nothing to settle."

Every winter there existed the threat of snowslides. A series of disastrous slides started in March of 1902. One hit the Liberty Bell sweeping away the aerial tramway and several men. A second slide buried a rescue party which was searching for bodies. The following day another drift forced a new group of rescuers to return to town. Meanwhile, a Cornish miner started the trek from ten miles east of Telluride to join the rescue effort. He was also buried by a snowslide. No bodies were uncovered for some time and it was months before the Cornish miner was found. Telluride was devastated by the chain of events.

The mines above Telluride produced gold, silver, copper, zinc and other minerals. For several years, silver was the backbone of Telluride's economy. Following the silver crash in 1893, the mining industry turned its attention to gold. Today, the economy thrives on tourism — and indeed its future is bright.

Through Placerville passes the famed Old Spanish Trail — route of the Dominguez-Escalante expedition in 1576.

PLACERVILLE
Location: 17 miles northwest of Telluride on State Highway 145

West of the Dallas Divide, along the San Miguel River, Colonel S. H. Baker and his group of prospectors discovered placer gold in 1876. A tent colony sprang up which was originally called Dry Diggings, then Hang Town. The following year its name was changed to Placerville. The Lower San Miguel Mining District was established as well.

The town of Placerville was platted in 1877. The original townsite was located at the intersection of present-day State Highway 62 and State Highway 145. Less than one half mile southeast of town, a general store and saloon were constructed on "Smith's Ranch" in 1879, and lots were sold. Through all of the '80s Placerville consisted of two sites. By 1890, when the Denver & Rio Grande Southern Railroad arrived at Placerville, most of the townsite had shifted, and the original location was soon to vanish.

The Bennett Dry Placer Amalgamator began operating in 1881. During the '80s the Keokuk Hydraulic Mining Company washed down an entire hill near Placerville. They maintained offices in the community, as did the St. Louis and San Miguel Company and the Mount Wilson Placer Mining Company.

As mining diminished, livestock became the backbone of Placerville. The Philadelphia Cattle Company established offices in town. As was the case in much of the old west, sheep herders and cattle ranchers didn't mix too well. On one occasion several sheepmen were killed in a range war. Ultimately Placerville became an important railroad shipping point for both cattle and sheep.

According to legend, Otto Mears was so terrified during his first ride over "his" Ophir Loop that he wanted to get out and walk.

OPHIR

Location: 14 miles south of Telluride via State Highway 145 and the Ophir Pass Road

Two miles east of State Highway 145 at the foot of Ophir Pass, lies the town of Ophir. At one time the town was called Old Ophir to avoid confusion with the community located at the Ophir Loop, two miles west — originally known as Howard's Fork, and later Ophir. If that's not confusing enough, the Ophir Station was also located at the loop, and the small community located there was sometimes referred to by that name.

Although there are other versions for the name's origin, Ophir was probably named for the biblical site where gold was discovered at King Solomon's Mine.

Lt. Howard, for whom Howard's Fork is named, made the first strike in the area in 1875. Many others followed. The north face of Yellow Mountain was where it all started. The Butterfly-Terrible, Silver Bell, Caribbean, Nevada, and the Badger were among the more profitable mines. And, they produced constantly right through the turn of the century.

Ophir (Old Ophir) had a population of about 400 at its peak. By the late 1890's the town had a water works, electricity, a hotel, a stamp mill, churches, and a school house — although a future school master was run-off for failing to join the miner's union.

Otto Mears was a road-builder and railroad builder who didn't know the word "impossible." The Ophir Loop was one of his crowning achievements. The major construction project enabled the railroad to run from Telluride to Durango. Three tiers of tracks with loops crossing above and below each other and trestles sometimes one hundred feet high were the result of Mears' challenge. Depending upon how they reacted, passengers were either thrilled or chilled by the experience.

The history of Ophir and Ophir Pass is sprinkled with tragedy — mostly due to the harsh winters and destructive avalanches. A story often told is that of Swedish mailman, Sven Nilson, who carried the mail back and forth from Silverton to Ophir throughout the year, even in the most hazardous weather. In fact, the winter trek over the pass was so dangerous he was paid a salary more than double that of the average miner. In December of 1883, Nilson left Silverton during a blizzard carrying Ophir's Christmas mail. He never reached Ophir. Several searchers combed the area, but to no avail. More than a year and a half later, Nilson's body was found in a ravine — with the pouch carrying Ophir's Christmas mail still strapped to his back.

The population of Ophir gradually declined after the turn of the century, until it had become a ghost-town by the 1920s. Today there has been a revitalization of Ophir, and the town is once again inhabited — mostly with summer residents.

ALTA

Location: South of Telluride; 5 miles east of State Highway 145 on the road to Alta Lakes

The Alta Mine was discovered in 1878, and about a mile away a cluster of cabins were built which became the town of Alta. The following year, in 1879, the famous Gold King Mine was located. The Gold King was a rich mine, but expensive to operate until L. L. Nunn brought in electrical power (see Ames).

The history of Alta followed that of the mines. During periods when they were closed, the town was virtually empty. The Gold King Mine worked more often than not, and did so through World War II.

Alta, which is situated in a very picturesque location, was never very large. In fact it never had a church or post office. However, it did have a school, a large boardinghouse, several mining buildings, an aerial tram, and many cabins. The entrance to the Black Hawk Tunnel is at Alta. It's 9,000 feet of passages reach both the Alta and St. Louis veins.

In 1945, a tragic incident occurred at the last of Alta's three mills. A fire raged while seven men were beneath the earth's surface. The superintendent ordered the portal dynamited in order to cut off the draft which was feeding the fire. One of the men sealed underground was the superintendent's son.

Due to the longevity of the Gold King Mine, Alta became a ghost-town later than most. As a result, many deteriorating buildings still stand.

Legend says L. L. Nunn became wealthy by charging miners 50 cents to bathe in the bathtub which he installed in his home.

AMES

Location: 15 miles south of Telluride via State Highway 145, near Ophir

The town of Ames sprang to life in 1882 when a smelter was constructed by the San Miguel River. Because of poor planning, the life of the smelting works was short lived. The San Miguel River ran through a deep canyon. Transporting ores into the canyon for smelting was an expensive proposition. Other smelters were built at more geographically feasible locations. When the smelter at Ames closed most of the town packed up and moved on.

It wasn't long before Ames experienced a revival. The Gold King Mine at Alta was experiencing difficulties because of its great costs of operating at such a high altitude — 12,000 feet. The Gold King and the town of Ames were saved by the visionary genius of L. L. Nunn, the mine's attorney.

Nunn envisioned building an electrical plant at Ames which would harness the power of the San Miguel River, then transmit it up to the Gold King Mine. Many scoffed at the idea but the plant was built anyway. It worked — fuel costs were cut and expenses were trimmed — the Gold King was saved. The result was the first commercial transmission of high-pressure electricity in history.

The success of the endeavor prompted Nunn to build high-tension lines across Imogene Pass (elevation: 13,000 feet) to serve Camp Bird and other mines on the Ouray side of the range.

PANDORA

Location: 1 mile east of Telluride

Below spectacular Bridal Veil Falls with its water cascading 365 feet down the cliffside, in an area once called Water Fall Gulch, lies the site of Pandora. Initially there were two small camps called Folsome and Newport at this location. Folsome preceded Newport, which was the original name of Pandora. The two little camps grew together to form the settlement which was named in 1881 for the Pandora Mine. The old Imogene Pass Road (which has been closed for years) originally ascended from Pandora across to Camp Bird.

The Smuggler-Union Mill was the main source of livelihood for the residents of Pandora. Aerial trams carted ore from the Smuggler and Union mines, high on the hill above town, down to the mill for processing. The Smuggler-Union had a history of labor strike and violence. In 1902, mine manager Arthur Collins was assassinated while sitting in his Pandora home. His replacement, B. Wells, narrowly escaped death when a bomb beneath his bed exploded and destroyed the house. Both the assassination and attempted assassination were blamed on the Western Federation of Miners Union.

Today the Smuggler-Union and Liberty Bell properties and the site of Pandora are all on property owned by the Idarado Mining Company. The Idarado is removing tailings and cleaning up the area. The trailer park at Pandora is also being removed. By the year 2011, law will allow development of this property which by then should be prime real estate.

"If the food at the camp is a little weak tastin', the coffee's probably strong enough to average it out."

TOMBOY

Location: 5 miles east of Telluride via Forest Route 869 (Tomboy Road)

Otis C. Thomas located the Tomboy Mine in 1880 on Savage Fork, high above Telluride. Tomboy was Thomas' nickname, but the claim was staked for George Rohwer. Because of its rather inaccessible location, there was very little activity in the area for several years.

Following the silver crash of 1893 many miners turned their attention to gold. Suddenly the Tomboy Gold Mines Company began producing handsomely. The small settlement of Savage Basin Camp blossomed to a population of about 900, and was renamed Tomboy.

Tomboy was very reliant upon Telluride as a supply center but it didn't have to rely on Telluride for "pleasure." Midway between Tomboy and the Smuggler Mine was a red-light district called "The Jungle" — a mixture of brothels, poker dens, and saloons.

Mine owners tried valiantly to clean up the area, but without much luck.

As with most camps that were adjacent to successful mines, Tomboy's history followed that of the Tomboy Mine and other mining properties in the vicinity. In 1894 other rich strikes were made nearby — the Columbia and the Japan. The Rothchilds of London purchased the Tomboy in 1897 for $2,000,000. The high mountain camp (which is located over 2,600 feet in elevation above Telluride) was booming. Mining operations declined after the turn of the century. The Tomboy Mine closed in 1927.

The remains of Tomboy are rapidly deteriorating, but there is still much to see. A trip to the old townsite is enriched with spectacular scenery.

Ox teams pull covered wagons of supplies down the main street of early Telluride. *Archives, University of Colorado at Boulder.*

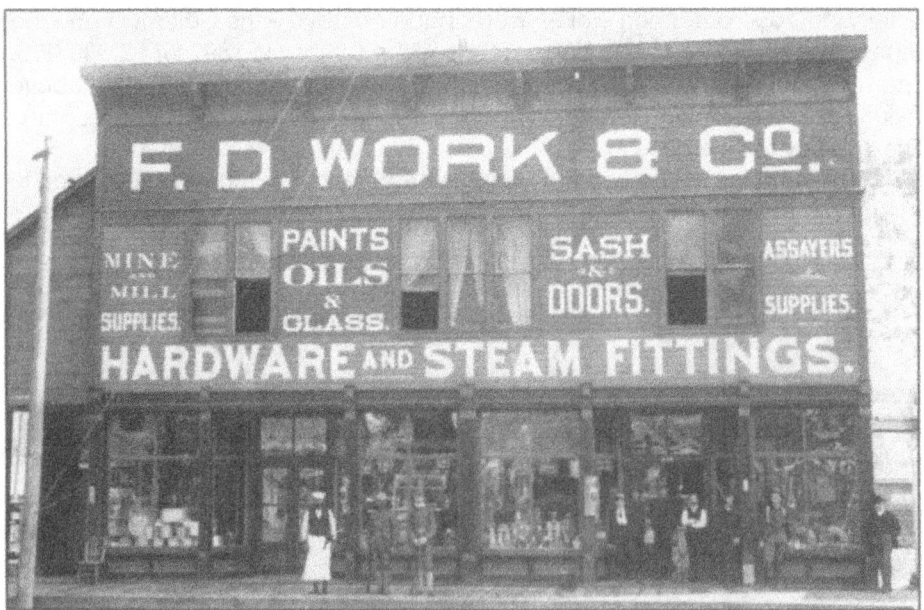

A storekeeper, militia men, and townspeople stand on a boardwalk during the labor strike at Telluride in 1903. *Denver Public Library, Western History Department.*

Looking down the main street of Ophir in August 1883. *Colorado Historical Society.*

The Gold King Mine at Alta. *Denver Public Library, Western History Department.*

The saloons at Ames empty for a photograph in 1883. *Colorado Historical Society.*

The Tomboy Gold Mines Company at Marshall Basin above Telluride. *Colorado Historical Society.*

DOLORES COUNTY

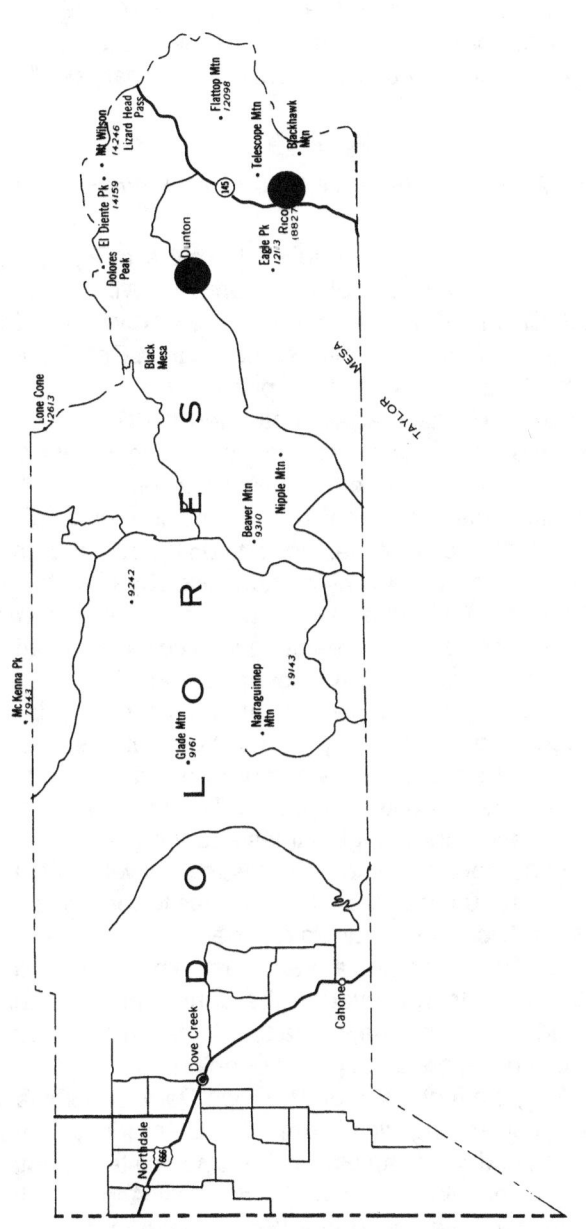

After a skeleton was discovered in 1882 by a fellow named Dave Munn the newspaper advised, "...some of the bones (are) now at the News' office; owner can have same by identifying and paying advertising charges..."

RICO

Location: 37 miles northeast of Dolores on State Highway 145

A rich strike was made here in July 1869 by a prospecting party from Santa Fe. The group was comprised of Joe Fearheller, William Hill, Jack Eccles, Tom Sager, "Tinker" Brown, "Pony" Whitmore, and a fellow named Miller. Several other strikes followed. Of the early mines, the Atlantic Cable and the Dolores (later to become the Aztec) were the most productive.

Although a smelting furnace was constructed in 1870, the camp remained a tent colony for several years just waiting for something to happen — and then it did. Senator John P. Jones and other Nevada investors purchased several lodes for $60,000 during the summer of 1879. The mill-sites were surveyed into lots. During the month of August alone, the first 105 log cabins were constructed within the community which was newly named Rico (Spanish for rich). The fall season of 1879 brought many other "firsts" to Rico. A fellow named "Frenchy" was murdered by "Kid" McGoldrick, resulting in Rico's first burial. McGoldrick was sentenced to six years which he never served. The first sermon was preached by Rev. H. P. Roberts of Silverton. After the sermon a large purse was collected for a horse race at the newly established Dolores Jockey Club, and everyone adjourned to the track. A postal route was established, an Episcopal church organized, and Rico had its first dance. The Thanksgiving Day Ball was held at Theodore Barlow's store. Eight ladies attended.

Though Rico had several saloons and was relatively wild, it also had its element of culture. In 1880 the Literary Society was formed. A school was also established with Miss Alice Snyder as its teacher.

There was a food shortage and a bread famine during the winter of '79-80. When John Foote's pack train arrived in the spring, he sold hundred pound sacks of flour for $35. With his new wealth, Foote built a dance hall which entertained a thousand visitors at its grand opening.

In January 1891, the first issue of *The Rico Democrat* advised, "The high grade mines, foremost among which are the Enterprise, Jumbo, C. H. C., Montezuma, Cobbler, and Newman Group, have had a steady output during the year past." More producing mines were the Golden Age, Aspen, Nebraska, Vestal, General Logan, Amazon, Stanley, Butler, Snow Flake and many others. The Enterprise was located and modestly operated by David Swickhimer. Operating capital was acquired when Laura, his wife, won $5,000 on a lottery ticket. The mine developed into a top silver producer. David Swickhimer was elected sheriff in November 1883 and later owned a saloon which was locally known as "Swick's Place." He sold the mine in 1891 for $1.25 million dollars and became president of the Rico State Bank.

When the railroad of Otto Mears arrived in 1891, there was a celebration that lasted for days. The ensuing boom was greater than anything Rico had seen in the past. By 1892 the city had a population of 5,000. There were two newspapers, many stores and hotels, 23 saloons, and a red light district which was three blocks long. The city was named county seat, and the attractive brick Dolores County courthouse was constructed.

The following article appeared November 21, 1891, in *The Rico Weekly Sun*. It might be representative of the male mind-set in Rico at the time. The article should be self explanatory — well, possibly. It read as follows:

"Rico may not be able to compete with neighboring towns in wrestling, or distant states in rock drilling, but we're well up on low-grade ore shipments, waltz quadrilles and beautiful women. Call at the Sun office and see our galaxy of female loveliness. The constellation embraces everything from carrot-tinted blondes to buxom brunettes."

The devaluation of silver in 1893 dealt Rico a severe blow. Many businesses closed and the population quickly dropped below 1,000.

The area mines have produced many minerals — and have continued to do so through the twentieth century. Although only a shadow of the old community, Rico endures today. During renovation in 1992 on the old Catholic Church (once the schoolhouse) the original old blackboards were uncovered.

DUNTON

Location: 10 miles west of State Highway 145, or 36 miles northeast of Dolores

Dunton sprang into existence in 1885 with the discovery of the Emma Mine. As other claims were established along the West Dolores River, the town blossomed. Dunton was a lively place with plenty of saloons to serve its nearly all-male population of three hundred.

According to legend, the infamous Butch Cassidy fled to Dunton after robbing a bank at Telluride.

Nearly half of Dunton's residents worked at the Emma. When the mines began to dwindle after the turn of the century, so did the population. The pastoral splendor of the isolated community was not enough to hold those who were miners by trade.

Buildings have been re-roofed and well preserved. Today Dunton attracts fishermen, hunters, hikers, and cross-country skiers.

Wagons stop opposite Mary Brown's Bon Ton Restaurant at Rico, where meals cost thirty-five cents. *Denver Public Library, Western History Department.*

La Plata in the mid-1880s. *Colorado Historical Society.*

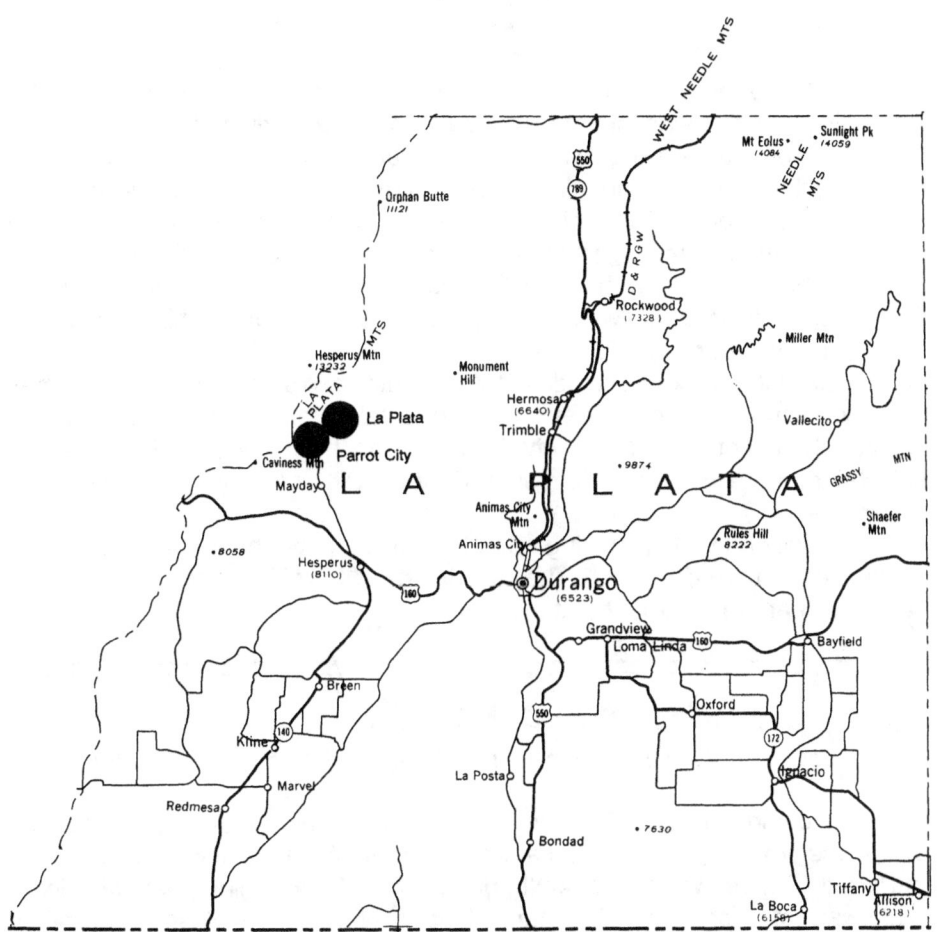

LA PLATA COUNTY

PARROTT CITY
Location: 17 miles northwest of Durango

In 1873, John Moss and his group of prospectors found gold along the La Plata River. As this was Indian territory, Moss made a private treaty with Chief Ignacio of the Southern Utes whereby he traded blankets and a hundred ponies for the right to prospect and mine an area thirty six square miles in size.

Moss, formerly a California miner, carried samples back to San Francisco and showed them to mining investor, Tiburcio Parrott. He agreed to stake Moss and sent him on his way back to Colorado. Although Parrott never lived in Colorado, the town which was established in the center of the thirty six square miles bore his name.

A sawmill was constructed in 1876. A hotel, two stores, and many homes followed. John Moss made a strong effort to have Parrott City selected as the first county seat of La Plata County. He failed and the seat went to Howardsville (which later became part of San Juan County). The seat eventually moved to Durango.

As Moss moved his mining operations further into the mountains, Parrott City became more deserted. The remaining old buildings were torn down by cattle ranchers who now inhabit the area.

LA PLATA
Location: 20 miles northwest of Durango via U.S. Highway 160

The community of La Plata was short-lived. Its birth was created because of its close proximity to the successful mines north of Parrott City. The population of Parrott City, below it to the south, peaked about 1880. The post office in La Plata was established in 1882. Many of the miners from the Moss camp shifted their residences closer to where they were working.

The area around La Plata, Spanish for silver, ironically produced more gold.

La Plata never had a railroad, but it was the northern most terminus of the stage line from Hesperus, a coal mining community south of Parrott City. The population of La Plata City (as it was originally called) peaked at about 200 in 1882. The post office was discontinued in December of 1885. By 1887, the population had dwindled to a mere 26.

Glossary of Mining Terms

ALLUVIAL DEPOSIT: Sediment deposited by a stream.

ARASTRA: An old Spanish apparatus used to break up ore by means of a heavy stone dragged around a circular trough.

ASSAY: To test and examine ores and minerals by a chemical process or the blowpipe method. Sometimes the assaying process requires the separation of precious metals from base metals by use of a cupel.

BAR: The peripheral accumulation of rocks along the banks of a stream, often worked for gold by prospectors.

CARBONATES: Those ores containing a large amount of carbonate of lead.

CLAIM: A mining claim is the right, or claim, of an individual or company to a specific location with set boundaries, and recorded according to law. Mining laws consider a claim legitimate if sufficient quantities of a metallic or other substance are found rendering the land valuable. Deposits of certain minerals, such as coal and oil, are the property of the United States and are not subject to claim under United States mining laws.

CONCENTRATION: The process by which nonessential and less valuable portions of ore are removed by mechanical means.

CRADLE: See Rocker.

CRIBBING: The process of constructing close timber, such as bulkheading or lining of a shaft.

CROSSCUT TUNNEL: A transverse tunnel which intersects a main tunnel, or drift, at an angle and leads to another point.

CUPEL: A small porous cup used in the assaying process of separate precious metals from lead and other base metals.

DREDGE: Bucket dredges and traction dredges are operated by power and usually are mounted on a boat. The bucket operates like an elevator and brings continuous loads of sand and gravel to the deck where it passes through a sluice. Traction dredges scoop much in the manner of a steam shovel. Suction dredges are smaller and often portable.

DRIFT: An underground tunnel which follows a vein. Usually the mine's main tunnel.

FISSURE VEIN: A crack in the earth's surface rock which is filled with a mineral matter other than that of its surrounding walls.

FLOAT: Pieces of ore which have washed away or have fallen from their parent veins. The discovery of float was usually the catalyst to trace each fragment toward its source.

GELENA: A common lead sulphide which often has silver content.

HYDRAULIC MINING: High pressure water is carried through a hose and nozzle and used to wash away gold bearing earth. The water and earth are carried through sluices which separate the gold.

IRON PYRITE: A mineral which resembles gold. Commonly called "fool's gold".

LODE: A vein or tabular deposit of precious mineral.

LONG TOM: A device used in the early mining days to aid in the separation of materials. Gravel is shoveled into the long tom, through which water is funneled, and worked with a hoe or rake. Gold and other heavier minerals are then swept through a screen and caught in the riffles beyond.

MILL: An establishment in which ores are reduced by means other than smelting.

MINE: Technically it is an ore deposit — a rich or abundant source. Commonly considered the pit, excavation, or tunnel from which ores and precious minerals, etc. are taken from the earth.

MINING DISTRICT: An area of country usually located within certain natural boundaries, and designated by name, in which a substantial amount of mining activity occurs.

MOTHER LODE: The predominant vein, or lode, passing through a particular area. Prospectors dreamed of discovering a mother lode.

NUGGET: A lump of native precious metal (i.e. a gold nugget).

ORE DEPOSIT: The primary source of the mineral which occurs in a vein.

PAN: The slowest method in searching for gold is panning. Water is swirled in a circular motion over the earth and gravel in a flat shallow pan. Sand and earth are gradually washed away.

PLACER MINING: Surface deposit mining, as placer mining is sometimes called, is one of the oldest methods. Water action has already extracted the precious material and deposited it in more accessible places to be worked, such as stream beds, etc.

QUARTZ: A common opaque mineral sometimes found near richer deposits.

REDUCTION WORKS: Any plant which reduces metal from its ore (i.e. a smelting works).

RIFFLE: The bottom of a sluice or trough with slats spaced closely together in order to catch gold and other heavy minerals.

ROCKER: An early mining device is the rocker, sometimes called a cradle. Earth and gravel are shoveled into a sieve box in the top of the rocker. Water is channelled over the sieve as the device is rocked to and fro. Heavier gold filters to a trough where it is caught by the riffles.

SAMPLING WORKS: An establishment in which ores are sampled to determine their value.

SLUICE: In areas where there was a good water supply, the sluice was one of the most popular devices used by miners. Dirt is shoveled into the long trough with transverse riffles. The water flow washes away waste material leaving gold and other heavy materials in the riffles.

SMELTING: A process by which metals are reduced from their ores by fusion, in a furnace or crucible.

STAMP MILL: An establishment or works where rock is crushed by steam-powered or water-powered pestles or stamps (i.e. a quartz mill).

TAILINGS: Residue which is left behind after precious metals have been separated from the ore by concentration or dressing.

TELLURIDE: A rich ore containing compounds of tellurium and gold and/or silver.

TRAMWAY: A cable system, suspended between two points, by which ore or other material may be transported by buckets.

VEIN: An elongated mineral deposit, or fissure, often rich in content.

VUG: An open cavity in a rock or formation which is sometimes lined with crystalline deposits.

WILFLEY TABLE: A table used in ore dressing for concentrating and seperating various metals. A jerking motion allows light grains to wash over a riffled surface while heavy grains remain. An important Colorado invention which increased recovery and profits.

WIRE GOLD OR SILVER: Native gold or silver in a maze of wire-like threads.

Acknowledgements

My sincere appreciation to the following:

Nancy Flanders, my sister, for her suggestions, and trips with me to visit many mining camps
Ruth S. Bennett, my mother, for proofreading with a critical and grammatical eye
Constant W. Southworth, my father, who first introduced me to Colorado mining camps in 1951
Teresa Bond, for her fine manuscript typing and long hours
Hal Flanders, my brother-in-law, for his help and suggestions, and for driving vehicle #2 on some mountain convoys — which included a little bad luck such as flat tires on Tin Cup Pass
Ray B. Walling, who logged about 7,000 miles stomping around mountain tops with me
Chip Southworth, my son, for his patience with this project
Dan Southworth, my son, for his graphics assistance
Herb and Sheri Gable, my friends, who couldn't handle the snow and run-off at Holy Cross City, but who accompanied me on visits to many mine and mining camps
Doreen Wollmer, for her excellent proofreading and suggestions
Sue Anderson, for her assistance and excellent grammatical skills.
Stephanie Bryant, my secretary, who was "glued" to a computer for hours and hours of transcribing
Luther E. Bennett, my stepfather, for his critique
Gene and Faye Franklin, my friends, who have visited many mining camps with me
Heather Flanders, my niece, who has "tagged-along" in some mighty high places
Becky Lintz and Barbara Foley, Colorado Historical Society, for a great amount of assistance during my research at the Stephen H. Hart Library
The Colorado History Museum
Frank, at the Granite Store, for his help and information
Julie and Doug in Crystal, for their hospitality, information, and a diet coke for this thirsty camper
Missy, Lisa, and Michelle Thompson, and their mother Joan, for accompanying me to many mining camps — even though the latter two hate mountain ledges
Dr. Stephen Eubanks of Denver for his assistance
Melanie Milam, whose family now owns the Stark Bros. Store in St. Elmo, for her help
Suzy Kelly, who seems to be "the" historian of Buena Vista
Ron Joe Hansen, The Mining Gallery, Leadville, for some interesting information

Leda Reed, of Tolland, who like this writer has an enthusiasm for ghost-towns
Marsala Hancock and Rilla Wiggins, Hahn's Peak Historical Society, for their "personalized" history
Henry Toll, owner of the towns of Baltimore and Hessie
Bob Glassman, of Baltimore, for his help and information
The self-proclaimed "Odd Fellow" of Russell Gulch, who resides in the old Odd Fellows building
William G. Kazel, Lake Gulch Milling Co., of Golden, for information on the Como Round House
Cassandra M. Volpe, University of Colorado, University Archives, for her suggestions and assistance
David M. Hays and Marilyn Burns, University of Colorado, Norlin Library, for their assistance
Mike, the Aspen Historical Society's resident "ghost" at Ashcroft
Carol Davis, South Park City Museum, Fairplay
The management of the Fairplay Hotel
Ramona Markalunas, Aspen Historical Society
Lisa Hancock Johnson, Aspen Historical Society
Bill Carter, another ghost-town author, for his assistance
Boyd Hill, who owns the oldest house in Jamestown
Rebecca Waugh, Summit Historical Society, Breckenridge, for her assistance
Robert E. Richardson, National Archives and Records Administration, Washington, D.C., for his cartographic assistance
Katie Verbois, who thinks you're rich when you pan out a few gold flakes
Frisco Historical Society
The Georgetown Society, Inc.
Christine Bradley, Archives (Georgetown), for her research assistance on Clear Creek County
The Gilpin County Historical Society
The Basalt Library
The Central City Opera House Association
Augie Mastrogiuseppe, Denver Public Library, Western History Department
Kathey Swan, Bruce Hanson, Phil Panum, Kay Wisnia, Mary Daze, Britt Kaur, and other staff members, Denver Public Library Western History Department
The Boulder County Metal Mining Association
The folks at the Rand Store in North Park
Creed Historical Society
Como Roundhouse Preservation, Inc.
The National Mining Hall of Fame & Museum, Leadville
Leadville - Lake County Chamber of Commerce
The Herald Democrat Newspaper Museum, Leadville
United States Geological Survey
Colorado School of Mines
Ed Hunter and Marty Miller, Cripple Creek-Victor Gold Mining Company

Terry Huber, Cripple Creek
The Buena Vista Chamber of Commerce
The Salida Museum
Crested Butte - Mt. Crested Butte Chamber of Commerce
Lake City Historical Museum
Lake City Chamber of Commerce
Buena Vista Heritage Museum
Museum of Western Colorado, Grand Junction
Randy and Margie Woods, for their verification and assistance regarding the site of Ruby
Bureau of Mines, Denver
The Historical Museum of Silverton
Kris Maxfield, The Silverton Chamber of Commerce
The Carnegie Library, Silverton
Allen Nossaman, Silverton
Animas Museum, Durango
Tread of Pioneers Museum, Steamboat Springs
The Colorado Railroad Museum, Golden
Glenn Campbell, who owns the old Ben Revett home in Breckenridge, for his hospitality and information
The Colorado Outfitters Association
The U. S. Forest Service - especially the district offices in Minturn, Aspen, and Salida
The George Rowe Museum, Silver Plume
Gary St. Clair, Mayor of Empire, for his valuable information
Fran Richardson of Empire for her gracious help
Doris Porth, Custer County Treasurer, for her research
Roxanne Eflin, Aspen, for her assistance
Grand County Museum, Hot Sulphur Springs
Cathy Gilbert, for her assistance and patience
Jeffrey Devene, Gilpin County Historical Society for his research
Silverton Chamber of Commerce
Marilyn Peterson, Colorado Mountain Club
Beverly Rich, San Juan County Historical Society, for her verification and assistance
Jim Wier, the trek master of Grand County
Jeani Speciale for her support and assistance
Bruce Montgomery, University Archives, University of Colorado at Boulder
Joel Sommers, for his assistance
David Halaas and David Wetzel, Colorado Historical Society
Marty Covey, University of Colorado at Boulder

Dave Southworth

Bibliography

BOOKS

Aldrich, John K. *My Favorite Ghosts.* Lakewood: Centennial Graphics, 1988.

Baker, James H., Editor. *History of Colorado.* Denver: Linderman Co., Inc., 1927.

Bancroft, Caroline. *Augusta Tabor: Her Side of the Scandal.* Boulder: Johnson Publishing Co., 1955.

Bancroft, Caroline. *Colorado's Lost Gold Mines and Buried Treasure.* Boulder: Johnson Publishing Co., 1961.

Bancroft, Caroline. *Silver Queen: The Fabulous Story of Baby Doe Tabor.* Boulder: Johnson Publishing Co., 1955.

Bancroft, Caroline. *Six Racy Madams of Colorado.* Boulder: Johnson Publishing Co., 1965.

Bancroft, Caroline. *Tabor's Matchless Mine and Lusty Leadville.* Boulder: Johnson Publishing Co., 1960.

Bancroft, Caroline. *Unique Ghost Towns and Mountain Spots.* Boulder: Johnson Publishing Co., 1961.

Barney, Libeus. *Letters of the Pike's Peak Gold Rush.* San Jose, CA: Talisman Press, 1959.

Baskin, O. L., & Co. *History of the Arkansas Valley, Colorado.* Chicago: O. L. Baskin & Co., 1881.

Bateman, Alan M. *Economic Mineral Deposits.* New York: John Wiley and Sons, Inc., 1958.

Bates, Margaret. *A Quick History of Lake City, Colorado.* Colorado Springs: Little London Press, 1973.

Bauer, William H., James L. Ozment and John H. Willard. *Colorado Post Offices: 1859 -1989.* Golden: The Colorado Railroad Museum, 1990.

Benham, Jack L. *Camp Bird and the Revenue.* Ouray: Bear Creek Publishing Co., 1980.

Bird, Allan G. *Bordellos of Blair Street.* Grand Rapids: The Other Shop, 1987.

Bird, Allan G. *Silverton Gold.* Grand Rapids: The Other Shop, 1986.

Bird, Allan G. *Silverton: Then and Now.* Englewood: Access Publishing, 1990.

Bishop, Isabella Bird. *A Lady's Life in the Rocky Mountains.* Norman: University of Oklahoma Press, 1960.

Blair, Edward. *Everybody Came to Leadville.* Leadville: Timberline Books, 1971.

Blair, Edward. *Leadville: Colorado's Magic City*. Boulder: Pruett Publishing Company, 1980.
Brown, Robert L. *Colorado Ghost Towns-Past and Present*. Caldwell: Caxton Printers, 1977.
Brown, Robert L. *Holy Cross - The Mountain and the City*. Caldwell: Caxton Printers, 1970.
Brown, Robert L. *Jeep Trails to Colorado Ghost Towns*. Caldwell: Caxton Printers, 1978.
Bueler, Gladys R. *Colorado's Colorful Characters*. Boulder: Pruett Publishing Co., 1981.
Carter, William. *Ghost Towns of the West*. Menlo Park, CA: Lane Publishing Co., 1971.
Clifton, Chas S. *Ghost Tales of Cripple Creek*. Colorado Springs: Little London Press, 1983.
Coquoz, Rene. *Tales of Early Leadville*. Leadville: Rene L. Coquoz Books, 1966.
Crofutt, George A. *Crofutt's Grip-Sack Guide of Colorado*. Boulder: Johnson Books, 1885.
Crossen, Forest. *The Switzerland Trail of America*. Boulder: Pruett Press, Inc., 1978.
Dallas, Sandra. *Colorado Ghost Towns and Mining Camps*. Norman: University of Oklahoma Press, 1985.
Dallas, Sandra. *Gaslights and Gingerbread*. Athens, OH: Swallow Press, 1965.
Dallas, Sandra. *No More than Five in a Bed: Colorado Hotels in the Old Days*. Norman: University of Oklahoma Press, 1967.
Dawson, John Frank. *Place Names In Colorado*. Denver: J. F. Dawson Publishing Co., 1954.
Dawson, Thomas F. and F.J.V. Skiff. *The Ute War*. Boulder: Johnson Publishing Co., 1980.
Decker, Sarah Platt (Durango Chapter D.A.R.) *Pioneers of the San Juan Country. Vol. I, II, III and IV*. Colorado Springs: The Out West Printing and Stationery Co., 1946 (1995 - new edition with indicies).
Dempsey, Stanley and James E. Fell, Jr. *Mining the Summit*. Norman: University of Oklahoma Press, 1986.
Dyer, J. L. *Snow-shoe Itinerant*. Cincinnati: Cranston & Stowe, 1890. Reprinted Breckenridge: Father Dyer United Methodist Church, 1975.
Eberhart, Perry. *Guide to the Colorado Ghost Towns and Mining Camps*. Denver: Sage Books, 1968.
Ellis, Amanda M. *Pioneers*. Colorado Springs: Dentan Publishing, 1955.
Ellis, Richard N. & Duane A. Smith. *Colorado: A History in Photographs*. Niwot: University Press of Colorado, 1991.
Fay, Abbott. *Famous Coloradans*. Ronia: Mountaintop Books, 1990.
Feitz, Leland. *Cripple Creek*. Colorado Springs: Little Long Press, 1967.

Feitz, Leland. *Ghost Towns of the Cripple Creek District*. Colorado Springs: Little London Press, 1974.

Feitz, Leland. *Soapy Smith's Creede*. Colorado Springs: Little London Press, 1973.

Fetter, Richard L. and Suzanne Fetter. *Telluride: From Pick to Powder*. Caldwell: Caxton Printers, 1979.

Field, Eugene. *A Little Book of Western Verse*. New York: Charles Scribner's Sons, 1894.

Florin, Lambert. *Ghost Towns of the West*. New York: Promontory Press, 1970.

Fossett, Frank. *Colorado*. New York: C.G. Crawford, 1880.

Gandy, Lewis Cass. *The Tabors*. New York: The Press of the Pioneers, Inc., 1934.

George, R.D. *Colorado Geological Survey*. Denver: The Smith-Brooks Printing Co., 1909.

Grimstad, Bill. *The Last Gold Rush*. Victor: Pollux Press, 1983.

Griswold, Don L. and Jean H. Griswold. *The Carbonate Camp Called Leadville*. Denver: University of Denver, 1951.

Hall, Frank. *History of the State of Colorado. 4 Vols*. Chicago: Blakely Printing Co., 1889, 1890, 1891, 1895.

Harrison, Louise C. *Empire and the Berthoud Pass*. Denver: Big Mountain Press, 1964.

Hollenback, Frank R. *Central City and Black Hawk*. Denver: Sage Books, 1961.

Hollon, E. Eugene. *The Lost Pathfinder - Zebulon Montgomery Pike*. Norman: University of Oklahoma Press, 1949.

Hollister, Ovando J. *The Mines of Colorado*. Springfield, MA: Samuel Bowles & Company, 1867.

Hunt, Inez and Wanetta W. Draper. *To Colorado's Restless Ghosts*. Denver: Sage Books, 1960.

Jessen, Kenneth. *Eccentric Colorado*. Boulder: Pruett Publishing Company, 1985.

Jocknick, Sidney. *Early Days on the Western Slope of Colorado*. Glorieta, NM: Rio Grande Press, 1968.

Lamm, Richard D. and Duane A. Smith. *Pioneers and Politicians*. Boulder: Pruett Publishing Company, 1984.

Larsh, Ed B. and Robert Nichols. *Leadville U.S.A*. Boulder: Johnson Books, 1993.

McCollum, Jr., Oscar. *Marble - A Town Built on Dreams. Volumes 1 & 2*. Denver: Sundance Publications, Ltd., 1992.

McKenney's Business Directory of Principal Towns in California, Nevada, Utah, Wyoming, Colorado and Nebraska 1882. San Francisco: H.S. Crocker & Co., Publishers, 1882.

McLean, Evalyn Walsh. *Father Struck It Rich*. Boston: Little, Brown and Co., 1936.

Marshall, John B. and Temple H. Cornelius. *Golden Treasures of San Juan.* Athens: Swallow Press, 1990.
May, Stephen. *Pilgrimage: A Journey Through Colorado's History and Culture.* Athens, OH: Swallow Press, 1987.
Monnett, John H. *Colorado Profiles: Men and Women Who Shaped the Centennial State.* Evergreen: Cordillera Press, 1987.
Montgomery, Mabel Guise. *A Story of Gold Hill, Colorado.* no pub., 1930.
Morris, John R. *Davis H. Waite: The Ideology of a Western Populist.* Washington, D.C.: University Press of America, 1982.
Mumey, Nolie, Editor. *Edward Dunsha Steele 1829 - 1865.* Boulder: Johnson Publishing Co., 1960.
Newton, Harry J. *Yellow Gold of Cripple Creek.* Denver: Nelson Publishing Co., 1928.
Noel, Thomas J. *Historical Atlas of Colorado.* Norman: University of Oklahoma Press, 1993.
Nossaman, Allen. *Many More Mountains. Volume 1: Silverton's Roots.* Sundance Books, 1989.
Olsen, Mary Ann. *The Silverton Story.* Cortez: Beaver Printing Co., 1962.
Perkin, Robert L. *The First Hundred Years.* Garden City, N.Y.: Doubleday & Co., Inc., 1959.
Perry, Eleanor. *I Remember Tin Cup.* Shawnee Mission, KS: Inter-Collegiate Press, Inc., 1986.
Poor, M.C. *Denver South Park & Pacific.* Denver: Rocky Mountain Railroad Club, 1976.
Pough, Frederick H. *A Field Guide to Rocks and Minerals.* Boston: Houghton Mifflin Company, 1955.
Prucha, Francis Paul. *American Indian Treaties: The History of a Political Anomaly.* Berkeley: University of California Press, 1994.
Rinehart, Frederick R. *Chronicles of Colorado.* Boulder: Roberts Rinehart, Inc., 1984.
Rockwell, Wilson. *Sunset Slope; True Epics of Western Colorado.* Denver: Big Mountain Press, 1956.
Schulze, Suzanne (Ed.). *A Century of the Colorado Census.* Greeley: University of Northern Colorado, 1976.
Shaputis, June and Suzanne Kelly (Ed.). *A History of Chaffee County.* Marceline, MO: Walsworth Publishing Co., 1982.
Skala, Helen and Dora Krocesky. *Leadville's Tales from the Old Timers.* Leadville: Skala and Krocesky, 1972.
Skala, Helen. Leadville's Tales From the Old Timers: Book 2. Leadville: Skala, 1977.
Smith, Duane A. *Rocky Mountain West.* Albuquerque: University of New Mexico Press, 1992.
Smith, P. David. *Ouray. Chief of the Utes.* Ouray: Wayfinder Press, 1986.
Sprague, Marshall. *Massacre.* Lincoln and London: University of Nebraska Press, 1980.

Stevenson, Thelma V. *Historic Hahns Peak*. Fort Collins: Robinson Press, 1979.
Ubbelohde, Carl. *A Colorado History*. Boulder: Pruett Press, Inc., 1965.
U.S. Bureau of the Census, Revised by the Social Science Research Council. *The Statistical History of the United States from Colonial Times to the Present*. Stanford: Fairfield Publishers, Inc., 1965.
Vandenbusche, Duane. *The Gunnison Country*. Gunnison: B & B Printers, 1980.
Vandenbusche, Duane, and Rex Myers. *Marble, Colorado: City of Stone*. Denver: Golden Bell Press, 1970.
Voynick, Stephen M. *The Making of a Hardrock Miner*. Berkeley: Howell-North, 1978.
Wallace, Betty. *Gunnison Country*. Denver: Sage Books, 1960.
Wentworth, Frank. *Aspen on the Roaring Fork*. Lakewood: Francis B. Rizzari, 1950.
Werner, Fred H. Meeker - *The Story of the Meeker Massacre and the Thornburgh Battle September 29, 1879*. Greeley: Werner Publications, 1985.
Weston, W., Editor. *The Denver, Northwestern and Pacific Railway*. Denver: The Denver, Northwestern and Pacific Railway, 1906.
Wolle, Muriel Sibell. *Stampede to Timberline*. Denver: Sage Books, 1962.
Wright, Carolyn and Clarence Wright. *Tiny Hinsdale of the Silvery San Juans*. Denver: Big Mountain Press, 1964.

NEWSPAPERS

Aspen Times
Boulder Daily Camera
Breckenridge Bulletin
Canon City Record
Carbonate Chronicle
Central City Daily
Central City Register-Call
Chaffee County Times (Buena Vista)
Clear Creek Courant (Idaho Springs)
Colorado Gambler (Denver)
Colorado Miner (Georgetown)
Colorado Prospector (Denver)
Creede Candle
Cripple Creek Gold Rush
Cripple Creek Times
Del Norte Prospector
Denver Post
Denver Republican
Denver Times
Denver Tribune
Dolores News (Rico)
Durango Herald
Durango Record
Durango Democrat
Empire Daily News
Fairplay Flume
Georgetown Courier
Gillett Forum
Grand Junction Daily Sentinel
Gunnison News

Gunnison Republican
La Plata Miner
Herald Democrat (Leadville)
Leadville Chronicle
Miner (Silverton)
Mineral County Miner (Creede)
Mountain Messenger (Idaho Springs)
Ouray Herald
Ouray Times
Park County Republican
Pine Cone (Apex)
Republican (Telluride)
Rico Democrat
Rico Sun
Rio Grande Magazine
Rocky Mountain News
San Juan Herald
Silver Standard (Silver Plume)
Silverton Democrat
Silverton Standard
Solid Muldoon (Ouray)
Steamboat Pilot
Summit County Journal (Breckenridge)
Summit County Leader (Breckenridge)
Telluride Journal
Wall Street Journal

ARTICLES

Anonymous. "The Latest Great Mining Camp." *Mining Industry and Tradesman II* (No. 23):229-37.

Ashby, Lindsey and Claude Wiatrowski. "Rebuilding of the Georgetown Loop Railroad." *Journal of the West*. (January 1992):60(10).

Bair, Everett. "A Journey to Old St. Elmo, A Glimpse into Her Boom Days." *1958 Grand Book*, 14:199-207. Boulder: Johnson Publishing Co., 1959.

Bond, Robert F. "South Park at Thirty." *Americana* (June 1989):10.

Bradley, Hassell. "Colorado Area Seen on Verge of New Gold Bonanza." *American Metal Market*. (December 13, 1990):4.

Colorado Historical Society. "James Peck, the Emperor of Empire." *Colorado Heritage*. (Summer 1990).

Colorado Historical Society. "Trails Through Time." *Colorado Heritage*. (Autumn 1990).

Duke, Ben. "A Journey Through Time." *Journal of the West*. (January 1990):67(9).

Economist, The. "One Good Hand Too Many." *The Economist*. (March 21, 1992):A28.

Engineering & Mining Journal. "Winfield Scott Stratton: 1848-1902." *Engineering & Mining Journal*. (May 1992):16LL.

Haase, Carl L. "Gothic, Colorado: City of Silver Wires." *Colorado Magazine* 51 (Fall 1974):294-316.

Halaas, David Fridtjof and Gerald C. Morton. "Boom and Bust: Images from the Colorado Chronicle." *Colorado Heritage.* Colorado Historical Society. (Issues 1 & 2, 1983).

Hastings, James K. "A Winter in the High Mountains, 1871-72." *Colorado Magazine* 27 (July 1950):225-33.

James, Louise Boyd. "The Case of the Colorado Cannibal or `Have a Friend for Dinner'." *American West.* (February 1990).

Jesitus, John. "Development in the Cards for some Colorado Towns." *Hotel & Motel Management.* (May 11, 1992):3(2).

Matthews, Carl F. "Rico, Colorado-Once a Roaring Camp." *Colorado Magazine* 28 (January 1951):37-49.

Naisbitt, John, Patricia Aburdene and John Vaughan. "West by Southwest: A Telluride Log House with Mining Camp Routes." *Architectural Digest* (June 1989):206(9).

Phelps, Richard W. "A Nugget in the Sangre de Cristos." *Engineering & Mining Journal.* (June 1992):45(3).

Smith, Duane A. and David Fridtjof Halaas. "A Fifty-Niner Miner: The Career of Horace W. Tabor." *Colorado Heritage.* Colorado Historical Society. (Issues 1 & 2, 1983).

Smith, Joseph Emerson. "Personal Recollections of Early Denver." *Colorado Magazine* 20, No. 2 (March 1943):62-64.

Spence, Clark. "Western Mining." *Historians and the American West.* Michael Malone, editor. Lincoln: University of Nebraska Press, 1983.

Spring, Agnes Wright. "Old Caribou and Central City." *1958 Brand Book*, 14:3-45. Boulder: Johnson Publishing Co., 1959.

Swanson, Evadene Burris. "Where's Manhattan?" *Colorado Magazine* 48 (February 1971):147-58.

Zavodni, Cathy. "Gold Mining Costs Keep Growing." *American Metal Market.* (June 17, 1991):A4.

Zimmerman, Karl. "Cliff-hanger." *Americana.* (May-June 1990):50(5).

Zintl, Amy. "Aspen, Colorado - Revival of a Once-Bust Boom Town." *Americana.* (February 1993):48(7).

Wolle, Muriel Sibell. "Irwin, a Ghost Town of the Elk Mountains." *Colorado Magazine* 24 (January 1947):4-15.

OTHER SOURCES

A Complete City Directory of Cripple Creek, Victor, and the Towns of the Cripple Creek Mining District. The Gazetteer Publishing Co., 1897.

American Metal Market - many brief articles too numerous to cite regarding the current Colorado mining scene and certain producers including Asarco, Inc., Galactic Resources, Ltd., Battle Mountain Gold Co., Cripple Creek and Victor Gold Mining Co., etc., with special recognition to authors Hassell Bradley and Christopher Munford.

Civil Works Administration, Thousand Town File, Unpublished, Colorado Historical Society, Denver.
Colorado Mining Directory and Buyers' Guide. Denver: G. A. Wahlgreen, 1901.
Cripple Creek District Directory, 1900. The Gazetteer Publishing Co., 1900.
Encyclopedia of American Business History and Biography: Railroads in the Age of Regulation, 1900-1980. New York: Bruccoli Clark Layman, Inc., 1988.
Encyclopedia of American Business History and Biography: Railroads of the Nineteenth Century. New York: Bruccoli Clark Layman, Inc., 1988.
First Annual Colorado Mining Directory, 1896. Compiled by J. S. Bartow and P. A. Simmons. Denver: The Colorado Mining Directory Co.
Leadville Mining District compiled from official records and other reliable sources, January 1901, Charles F. Saunders. Also copyright 1901 by Charles F. Saunders.
Map of Colorado Territory, Embracing the Central Gold Region. Drawn by Frederick J. Ebert under direction of the Governor, Wm. Gilpin. Published by Jacob Monk, 1862.
Map of Public Surveys in Colorado Territory. Map to accompany report of the Surveyor General, 1866, (Issued by the General Land Office on Oct. 2, 1866.)
Map of the State of Colorado. Compiled from the official Records of the General Land Office, 1987. Compiled by A. F. Dinsmore. Revised and corrected for reissue by M. Hendges.
Map of the State of Colorado, 1885. Compiled from the official Records of the General Land Office, Compiled and drawn by M. Hendges.
Map of the Territory of Colorado Showing the Extent of the Public Surveys. Map to accompany the Annual Report for 1871. Compiled under the direction of the Surveyor General.
Maps showing Topography and Mining Claims of the Alma District Colorado, 1912, Colorado State Geological Survey, R. D. George, State Geologist.
National Archives, Cartographic and Architectural Branch, Washington, D.C.
Nell's New Topographical and Township Map of the State of Colorado. Compiled from U.S. Government Surveys and other authentic Sources, 1881.
Nell's Map of Colorado, 1885. Chain and Hardy Co., Agent, Denver, 1885.
Nell's Map of Colorado, 1902. Hamilton and Kendrick, Agent, Denver, 1902.
Post Route Map of the State of Colorado showing post offices with the intermediate distances and mail routes in operation on the 1st of October, 1885.
Thayer's Map of Colorado Published by H. L. Thayer, Denver, Col., 1880. From Surveys of the General Land Office, used by permission, revised and corrected to date by the Publisher.

This is Climax Molybdenum Company: 1979.

United States Geological Survey, Maps, U.S. Department of the Interior, Federal Center, Denver.

United States Department of Agriculture, Forest Service Maps, U.S. Forest Service, Denver.

Williams' Tourist's Map of Colorado and the San Juan Mines. Engraved from Surveys by the Hayden U.S. Geological Expedition. Henry W. Troy, Designer, N.Y., 1877.

Index

A

A. Y. Mine 98
A.P. Stevens' Mill 43
Abbeyville 258
Adams, Alva (Governor) 56, 99, 162
Adams, Samuel (Captain) 178
ADELAIDE 102-103
Adelaide Mine 102
Aikins, Thomas (Captain) 53
Ajax Mine 113
Alamosa River 162
ALICE 44-45
Alice Mine 44
Alie Bell Mine 135
ALMA 78, 83, 91
Alma Station (Alma Junction) 82-83
ALPINE 134, 136, 143
ALPINE STATION 135, 257-258
Alpine Tunnel 135-137, 253-254, 257
ALTA 283
Alta Mine 283
ALTMAN 118-119, 126
America the Beautiful 111
AMERICAN CITY 24
American Girl Mine 243
American Mine 59
American Nettie Mine 240
American Sisters Group (originally Seven Sisters) 43
American Smelting & Refining Company 102
American Tunnel Mine 230, 232
American Tunnel Project 234
AMES 283-284, 288
Amethyst Mine 171
Ammons, Teller (Governor) 35
ANACONDA 120, 126
Anaconda Tunnel 120
Andracich, Frank 150
"Angel of Shavano" 133-134
ANIMAS FORKS 231, 233-234, 237, 273-275
Animas Forks Pioneer 233
Animas River 232, 234
Animas Valley 228
Anteso Hotel 130
Anthony, Susan B. 99, 272
APEX 23, 24
Apex Pine Cone 23
ARBOURVILLE 133, 138
Ardourel, Francois 60
Arequa 117
ARGENTINE 184-185, 196
Argentine Central Railroad 37
Argentine Mine 279
Argentine Pass 37, 183, 185
Argentum 275
Argo Mill 32
Argo Tunnel 32
Arkansas Mountain 60
Arkansas River 11–12, 129, 137
Arkansas Valley Plant 102
ARROW 224-225
Arthur, Chester A. (President) 221
ASHCROFT 199-201, 204, 259
ASPEN 104, 106, 200-204, 207, 218, 259, 265
Aspen Daily Times 202
Aspen Historical Society 199, 200
Aspen Junction 212
Aspen Mine 201, 202
Aspen Mountain 201-202, 207
Aspen Times (Aspen Daily Times) 203
Aspen Union Era 203
Astor City 210
Atlantic 257
Atlantic Mine 36
Atlas Mine 241

August Belmont Mine 38
Auraria 11
Aztec Mine 161

B

BABCOCK 131
BACHELOR 169-171, 173
Bachelor City Dramatic Club 169
Bachelor Mine 240
Bachelor Mountain 169
Badger Mine 282
Baker, Charles 228
Baker, S. H. (Colonel) 281
Baker's Park 228
Balarat 55
Bald Mountain Mining Company 252
BALDWIN 265
BALFOUR 86, 93
Balfour News Weekly 86
Ball, David J. 33
Ball Mill 182
BALTIMORE 22, 23
Baltimore Club 23
Baltimore Ridge 22
Bancroft, Caroline 98
Bank of Telluride 280
Banker Mine 140
Bard, Richard (Dr.) 33
Barnard, A. W. 39
Barton House 39
BASALT 212, 216
Bassick, Edmund C. 150
Bassick Mine 150
Bassickville 150
Bates, Katherine Lee 111
Battle Mountain (Teller County) 113, 115
Battle Mountain (Eagle County) 210, 213
Beacon Hill 117
Bear Mountain 262
Beaumont Hotel 240, 246
BEAVER CITY 139
Bebee, F. W. 31

Bebee House 31
Beeson, C. M. 230
Begole, A. W. 239
Belden, D. D. (Judge) 213
Belden Mine 213
Bell, J. W. (Dr.) 152
Bell, John 177
Belle of the West Mine 271
Belvedere Theater 17
Bennett Dry Placer Amalgamator 281
Bennett, Horace 111
Bennett, Thomas M. 77
Bergh House 77
Berta, Joseph "Peanuts" 265
Berthoud Pass 35
Biedell, Mark 156
Big Mine 259
Big Spar Mine 172
Billing and Eilers Smelter 102
Bishop, Jack 35
Black Canyon 22
Black Cloud Mine 63
BLACK HAWK 12, 16, 19-20, 26
Black Hawk Tunnel 283
Black Prince Mine 103
Black Queen Mine 262
Black Wonder Mill 275, 277
Black Wonder Mine 274
Blair, Tom 228
Blake, Otto 165
Blanca Mutual Mining and Milling Company 157
Blanca Peak 167
Blanton, William 42
Bloom, Sol 250
Blue Boarding House 36
Blue River 177, 178, 180
Blue River Methodist Mission 181
Bluebird 62
Bluebird Lodge 54
Bobtail Lode 19
Boettcher, Charles 98
BONANZA 156, 159

Bonanza King Mine 276
Bonanza Mine 156
Bonanza News 156
Bonhomme Mine 275
Bonito 44
Booth, Edwin 17
BOREAS 187-188
Boreas Pass 84, 185, 187-188
Boston and Colorado Smelting Works 12, 19, 27, 83
Boston Conservatory of Music 23
Boston Mining Company 182
Boston Silver Company 182
Boston Silver Mining Association 182
Boston-Occidental Mining Company 24
Boulder Canyon 53
Boulder County 53-69
Boulder County Map 52
Boulder County Mine 57
Boulder Creek 53, 62-63
Bowen, Thomas (Judge) 77
Bowen, Tom 161
BOWERMAN 266
Bowerman, J. C. 266
Brainard Camp (Camp Tolcott) 64
Brainard Mine 64
Brainard, Wesley (Colonel) 64
BRECKENRIDGE 84, 177-181, 187-192
Breckinridge, John C. (Vice-President) 177
Breed and Cutter Silver Reduction Works 58
Bridal Veil Falls 284
Bridger, Jim 261
Briggs, J. L. 228
"Broken Nose" Scotty 99
Bromley (see Brumley)
Brown, Clara 18
Brown, David R. C. 201
Brown Gulch 41
Brown, J. J. 98

Brown, James M. (Ten Die) 229
Brown, Margaret Tobin 98
Brown, N. W. 58
Brown, Robert L. 276
Brownsville 40-41
BRUMLEY 106
Brunot Agreement 228, 271, 273-274
Brush Creek 214
Brush, Fred 136
Buchanan, James (President) 177
Buckeye 33
Buckeye Peak 105
BUCKSKIN JOE 78-79, 83, 188
Bucktown 102
Buckwalter, Harry 122
BUENA VISTA 131, 137-138, 144, 201
Buena Vista Heritage Museum 138
Buffalo Springs 78
Buford, N. B. (General) 186
Bugtown 218
Burnett, Benjamin Franklin 222, 224
Burns, James 115
Burrows, Abe 231
Burrows Mine 231
Burrows Park 275
Butte Opera House 112
Butterfly-Terrible Mine 282

C

C. F. Abbey Smelter 258
CACHE CREEK 129
Cache Creek Mining Company 129
Cache La Poudre River 71
Cain, Tom (Marshal) 230
Calamity Jane 172
Calderwood, "King" 119
Caledonia Mine 22
California Gulch 96-97
California Mining District 97
Calumet 141
Calvin, C. L. 169
CAMERON 118, 125

Cameron Mines Land and Tunnel Company 147
CAMP BIRD 230, 244-245, 248, 284
Camp Bird Mine 105, 234
Camp Enterprise 55
Camp Fancy 211
Camp Providence 55
CAMP TOLCOTT (Camp Talcott) 64
Campbell, E. N. (Sheriff) 271
Campion, John F. 99, 104
Campion's Lodge 104
Canyon Creek 239
CAPITOL CITY 273-274
Carbonate National Bank 100
Carbonateville 188
Carbondale 262
CARDINAL 57-58, 62, 66
Cardinal City 57-58
Caribbean Mine 282
CARIBOU 56-58, 66, 187
Caribou Mine 17, 56, 58
Carlos, Juan 167
Carns, Harry 35
Carpenter, Cass 97
Carrow, Otto W. 56
CARSON 271, 276
Carson, Christopher J. 276
Carter Group of Mines, The 254
Caryl, Charles W. 61
Casey, Pat 18
Casey's Table d'Hote 53
Cashier Mine 36, 43
Cassidy, Butch (Robert L. Parker) 218, 280, 291
Castle Creek 200, 202
Castle Mine 201
Castleton 265
Cat Gulch 141
Cataract Creek 275
Cave Mine 214
Caverna del Oro 151
Cement Creek 228, 234
CENTRAL CITY 11–12, 16-20, 23, 25, 32, 34, 45, 56, 62, 100, 179

Central City Opera House 17
Central Region 75
Chaffee City 132
Chaffee County 129-145
Chaffee County Map 128
Chaffee, Jerome B. (Senator) 57, 132
Chalk Creek 137
Chalk Creek Pass 131
Champion Mine 275
CHANCE 158, 264
Charles Bermister's Park House 34
Cherokee Gulch 41
Chester Arthur (President) 98
Chicago Creek 31
Chief Ignacio 294
Chief Ouray 12, 199, 228, 239, 272
CHIHUAHUA 184-185
Chipeta 199
Chrysolite Mine 98
Cimarron Mine 279
Cinnamon Pass 274
Civil War 34, 178, 228
Clarence 263
Clarendon Hotel 100, 108
Clark, Jim (Marshal) 280
Clear Creek 31, 38, 129, 140
Clear Creek County 31-51
Clear Creek County Map 30
Clear Creek Gulch 140
Cleary, Harry 229
Cleveholm 206, 208
Cleveland, Grover (President) 101
Cleveland Mine 279
Clifton Hotel 136
Climax Molybdenum Company 189-190
Coal Basin 206
Coan, A. S. 60
Coburn, John 42
Cochetopa National Forest 140
Cody, William F. (Colonel) (Buffalo Bill) 17, 32, 99, 230

Coley, John 182-183
Colfax, Schuyler 80
Collier, D. C. 183
Collier Mountain 183
Collins, Arthur 284
Collins, Helen (Mrs.) 262
Colona Bar 186
Colorado Map 13
Colorado & Northwestern Railway (Switzerland Trail Line) 55, 58, 62
Colorado & Southern Railway 40, 185
Colorado Central Railroad 19, 39-40, 50, 105
Colorado Fuel and Iron Company 167, 206, 259
Colorado Midland Railway 202, 212
Colorado Miner (Georgetown) 39
Colorado National Bank 263
Colorado Prince Mine 103
Colorado Springs 111-112, 115, 119
Colorado Springs and Cripple Creek District Railway 112, 114-115, 117
Colorado State Fish Hatchery 254
Colorado State Museum 263
Colorado Yule Marble Company 263
Columbia Mine 55, 279, 285
Columbia Mountain 41, 43
COLUMBINE 219
Columbine Hotel 61
COMO 84-85, 185, 187
Como House 84
Como Roundhouse Preservation, Inc. 84
Compromise Mine 202
Compromise Mining Company 201
Comstock Mine (Virginia City, Nevada) 44, 56
Conejos County 164-165
Conejos County Map 163
Conejos River 164
CONGER CAMP 187

Conger, Sam P. (Colonel) 56-57, 187
Connors, Bill 229
Conqueror Mine 36
Conqueror Mining and Milling Company 37
Conqueror Tunnel 36
Coombs, J. H. 34
Coon Trail 57
Corbett, James 100
Cornwall, John 162
Cornwall Mountain 162
Corydon Rose 274
Costilla County 167
Costilla County Map 166
Cottonwood 157
Cottonwood Creek 131, 275
Cottonwood Pass 131
Covode Mountain 36, 43
Cowenhoven, H. P. 201
Cowles, Henry DeWitt Clinton (Judge) 34, 36
Cracker Jack Mine 275
CREEDE 99, 169-173
Creede Candle 170-172
Creede Mining District 169, 171
Creede, Nicholas C. 132, 171
Cree's Camp 131
Cresson Mine 117
CRESTED BUTTE 258-259, 262-263, 268
Crested Butte Republican 259
CRESTONE 157-158
Crestone Needle 157
CRIPPLE CREEK 12, 19, 62, 86-87, 111-115, 121-122
Cripple Creek Area Map 110
Cripple Creek Mining District 112, 117-119
Cripple Creek-Victor Gold Mining Company 118
CRISMAN 60, 68
Crisman, Obed 60
Crofutt, George A. 233

Crofutt's Grip-Sack Guide of Colorado 219, 233
CRYSTAL 261-263, 269
Crystal Canyon 261-262
Crystal Club 262
Crystal Mill, The 262
Crystal Mountain 262
Crystal River 206, 208, 262-263
Crystal River Current 262
Culver, C. B. 200
Cumberland Pass 252
Cunningham Gulch 232
Currant Creek Pass 86
Custer County 149-154
Custer County Map 148

D

Dailey, James 43
Darley, George M. (Rev.) 271
Davenport, Washington 219
Davis, Cyrus 234
Davis, Herndon 17
Davis, William (Rev.) 244
Day, David Frakes 239
Dayton 103
De Lamar, Joseph Raphael 150
Deardorff, Cy 55
Dearheimer, Alsina 137
Decatur 184-185
Decatur, Stephen 184
Del Norte 156, 161
DELAWARE FLATS 180
Dempsey, Jack 112
Denver 84, 87, 162, 180, 203, 250, 261, 263
Denver & Rio Grande Railroad 137, 141, 151, 167, 171, 186, 196, 200, 202, 204-205, 210-211, 213, 229, 242, 271
Denver & Rio Grande Southern Railroad 281
Denver & Salt Lake Railway 224
Denver and South Park Railroad 86, 134
Denver City 11
Denver City Mine 158
Denver Hotel 178, 190
Denver, Leadville & Gunnison Railway 92
Denver Mint 37
Denver Post 22, 272
Denver, South Park and Hill Top Railway 86
Denver, South Park & Pacific Railroad 84, 87, 91, 135-137, 178, 185-187, 254, 257-258
Devil's Gate High Bridge 40
DeWalt, Frank W. 99
Dexter, James Viola 99, 108
Dianthe Mine 187
Dives Mine 34, 41
Dodge City Cowboy Band (originally spelled Cow Boy) 101, 230
Doe, Jr., William Harvey 19
Dolores County Pgs. 290-292
Dolores County (Map) 289
Dominguez-Escalante expedition 281
DORCHESTER 265-266, 269
Dougan, David H. (Dr.) 99
Douglas County 73-74
Douglas County Map 72
Douglas Mill 22
Douglas, Stephen A. 177
Doyle, James 115
Doyle, William 218
Dry Creek 11-12
Dudley 78
Duggan, Martin 100
DUMONT 32-33
Dumont, John M. 32, 33, 36, 44
Dun, R. G. 57
Duncan 157
DUNTON 291
DuPuy, Louis 39
Durango 282, 294
Durango & Silverton Narrow Gauge Railroad Company 231

Durant Mine 201
Dutchtown 224
Dyer, Elias (Judge) 130
Dyer, John L. (Father) 79,
   99, 130, 136, 181, 183, 188
DYERSVILLE 187-188

E

E. C. Stoddard Company 252
Eagle County 210-216
Eagle County Map 209
Eagle Mine 213
Eagle River 210, 212
Eagle Valley Shaft (Redcliff) 210
Early Bird Mine 233
Earp, Wyatt 230
East Portal 22
East Willow Creek 171
Eaton, G. C. 262
Echo Bay Mines 232
Eckles, Jack 239
Eclipse 117
Edwards, R. J. 151
ELDORA 62-63, 69
Elizabethtown 38
Elk Mountain Hotel 259
Elkhorn 71
Elko 261
ELKTON 117-118
Elkton Mine 117
Ellis, Anne 156
EMMA 205
Emma Mine 201, 291
EMPIRE 33-36, 38, 43, 48
Empire City Mining Agency 34
Empire Ditch and Placer Company 36
Empire Zinc Company of New Jersey 213
Empress Josephine Mine 156
Engineer Pass 231, 273, 274
Engineer Peak 273
Enterprise Mine 290
Epsey, Bob 182

Eskridge, Dyson 230
Esmund, John 161
Esmund Mine 161
EUREKA 232-233, 235-236
Eureka Mountain 36
Evansville 103
Evening Chronicle 203
EVERETT 106
Everett, C. M. 106, 205
Exchequer Mine 156
Exchequerville 156

F

"Face on the Barroom Floor" 17
FAIRPLAY 77-78, 82, 84, 89
Fairplay Flume 84
Fairplay Hotel 77, 78
Fairview 254
Fairview Mine 253
Fall River 32, 44
Fallon, John 279
Farncomb, Harry 180
Farnham, Thomas Jefferson 79
Farwell Mine 199
Father Struck It Rich 234, 244
Feathers, Amasa 138
Federal Union Mine 186
Feenan, Owen 41
Fenton, James 99
Ferdinand the Seventh (King of
   Spain) 156
Field, Eugene 53
Finntown 97
First National Bank of Leadville 99
Fisher, Kate 250
Flora Bell Mine 135
Florence and Cripple Creek Railroad
   112, 114-115, 117, 120
Folsome 284
Fools Peak Trail 214
Foote, John 290
Ford, Bob 171, 172
Forest City 136
Forest King Mine 164

Forest King Mountain 164
Forest Queen Mine 260
Foresters 20
Fossett's *Colorado* 22
Fourmile Creek 60, 85
Fourth of July Mine 63
Free America 43
Free Coinage Hotel 172
Free Coinage Mine 61
Free Gold 137
FREELAND 33, 44, 51
Freeland Mine 32, 44
Freeman, Edgar 36
Fremont 111
Fremont County 147
Fremont County Map 146
Fremont, John 81, 157
Fremont Pass 105, 189
French Gulch 180, 181, 194
Freshwater (Guffey) 86
Freshwater Creek 87
Freshwater Mining District 87
FRISCO 186
Frisco Historic Park 186
Frisco Historical Society 186
Fryer, George 99
Frying Pan River 212
FULFORD 213-214
Fulford, Arthur H. 214
Fulford Cave, The 214
Fullen, Hiram 59, 62

G

Gabbert, William (Supreme Court Chief Justice) 117
Gage, Lyman J. 64
Galactic Resources, Ltd. 161
Galena 261
Galena Mountain 261
Galet, Sophie 39
Galina Gulch 256
Gambell, A. D. 21
Gambell Gulch 24
Garfield 138

Garo 78
Garrison, Emma Davis 205
Gaskill 221
Geary, Peter 34
Geneva Mining District 88
George's Town 38
GEORGETOWN 31-32, 38-42, 45, 49
Georgetown Courier 33, 39
Georgetown Loop 41
Georgetown Loop Railroad 40
Georgetown Miner 43
Georgetown Society 40
Georgia Bar 129
Georgia Gulch 83, 179
Georgia Pass 83, 84, 179
German Flats 254
Gerry, M. B. (Judge) 272
Gibbs, Elijah 130
Gibson Hill 181
Giles Mine 64
Gilheany, Jack 229
Gillespie, Henry B. 201-202
GILLETT 116, 124
Gillett Bull Ring 116
Gillett Reduction Works 116
GILMAN 212-213
Gilman Enterprise, The 213
Gilman, Henry M. 213
GILPIN 24
Gilpin County 16-29
Gilpin County Map 15
Gilpin, William (Governor) 11–12, 24, 157
Gimlett, Frank 133
Glacier Mountain 190
GLADSTONE 234-235, 237
Gladstone Kibosh 234
Glenwood Springs 202
Goddard, Luther (Justice) 117
Gold Anchor 45
Gold Belt, The 141
Gold Bug Mine 141
Gold Coin Club 114

Gold Coin Mill 123
Gold Coin Mine 113, 147
Gold Creek 254
Gold Cup Mine 250, 252
Gold Dirt 21, 24
GOLD HILL 53-54, 59-62, 65
Gold Hill Inn 54
Gold Hill Mining District 53
Gold King Mill 237
Gold King Mine (San Juan County) 234
Gold King Mine (San Miguel County) 283-284
Gold King Mine (Teller County) 111
Gold King Tunnel 234
Gold Link Mining Company 255
Gold Miner Hotel 63
Gold Mining Stock Exchange 111
GOLD PARK 211
Gold Park Mining Company 211, 212
Gold Prince Mill 233
Gold Run 53, 129
Gold Run Gulch 181
Golden 31
Golden Age Mine 54
Golden Circle Railroad 115
Golden Crescent, The (Cameron) 118
Golden Eagle, The (Duncan) 157
Golden Fleece Mine 271-272
Golden, Tom 31
GOLDFIELD 115-116, 124
Goldfield Leader, The 116
Goldfield Times, The 116
Goodwin, Charles 273
Gorman, A. Y. 96
GOTHIC 260-261
Gothic Mountain 260
Gothic Silver Record 260
Graham, Jack 224-225
Grand Central Hotel 178
Grand County 224-225
Grand County Map 223

Grand Imperial Hotel 229
Grand Island 63
Grand Opera House 112
Grand Republic Mine 60
Granger, Ralph 171
GRANITE 129-130, 139, 142
Granite Falls 271
Granite Mountain 257
Grant, Ulysses S. (President) 17, 32, 57, 156, 259-261
Gray, Ben 250
Gray, Charlie 250
Great Western Auction House and Clothing Store 98
Greeley, Horace 17
Green Mountain 141
Gregory Gulch 16, 19
Gregory, John H. 11-12, 16
Gregory Lode 11-12, 20, 25-26
Gresham 55
Griffin, Clifford 41
Griffith, David 38
Griffith, Elizabeth 38
Griffith, George 38
Griffith Mining District 38
Ground Hog Mine 213
Groves, Tom 181, 194
GUFFEY 86-87
Guggenheim, Meyer 96, 98, 230
Guinan, Mary Louise (Texas "Hello, Sucker" Guinan) 120
Gulf and Western Corporation 213
Gunnell Hill 17
Gunnison 254, 259, 264, 272
Gunnison County 250-269
Gunnison County Map 249
Gunnison River 275
GUSTON 242-244
Guston Claim 244

H

Hagerman Tunnel 202
Hahn, Joseph 218
HAHNS PEAK 216, 218-219

Halfmoon Gulch 106
Hall, Frank 279
Hall, J. M. 184
Hall Valley Mining District 87, 88
Hamill House 39
Hamill, Priscilla 39
Hamill, William A. 39, 41
HAMILTON 82
Hamilton Mine 182
HANCOCK 135-136, 145, 254
Hancock Placer 135
Handcart Gulch 87
Happy Valley Placer 62
Hardscrabble Mining District 149
Harnan, John 115
Harrison, George W. 16
Hartville 132
HARVARD CITY 131
Hatfield, John 23
Hatfield, Lillian Woodard 23
Hayden Placer 111
Haywood Hot Springs 130
Healy House Museum 99
Hendricks Mining Company 57
HENSON 271-273
Henson Creek 274
Henson, Henry 273
Hesperus 294
HESSIE 63
Hickok, Wild Bill 260
Hidden Treasure Mine 273
Higgenbottom, Joseph 78
High Line 112, 114
HIGHLAND 200, 204
Highland Town Company 204
Highland Tunnel 204
Hill, Nathaniel P. (Professor) 12, 19
Hill Top Mill 86
Hill Top Mine 85-86
Hiller, Edward 258
HILLERTON 251, 258
Hillerton Occident 258
Himrod, Peter 46
Hinsdale County 271-277

Hinsdale County Map 270
Hitchcock, Hiram 59
Holliday, John Henry "Doc" 32, 99
HOLY CROSS CITY 211-212, 215
Holy Cross Mining District 211
Holy Moses Mine 171
Homestake Creek 211
Homestake Mining District 104
Hook, George 98
Hooper, J. D. 201
Hoosier Pass 79, 82
Hoover, John J. 77
Hope Diamond 245
Hope Mining, Milling and Leasing Co. 204
Hord, Carl 250
Horn Peak 152
Horn, Tom 219
HORSESHOE 85
Horseshoe Gulch 85, 181
Horseshoe Mountain 85
Horsfel, David 53
Horsfel Mountain 53-54
HORTENSE 130
Hortense Mine 130
Hot Sulfur Springs 218
Hotchkiss, Enos 271
Hotel Arlington 178
Hotel de Paris 39
Hotel Jerome 202-203
Howard, George W. 232
Howard's Fork 282
HOWARDSVILLE 232, 294
Hoy, Valentine 218
Huff, James 38
Hull City Mine 117, 125
Humboldt Mine 149
Hunkydory Gulch 133
Hunley's Addition 200
Hunter, Gallant V. 34
Hunter, Luke (Sheriff) 230
Hyman, David 201

I

Ibex Camp 102
Ibex Mine 102
Ibex Mining Co. 102
Idaho Bar 31
IDAHO SPRINGS 11-12, 31-32, 45, 47, 186
Idarado Mining Company 279, 284
Iliff, William H. 177
ILSE 150-151
Imogene Basin 245
Imogene Pass 284
Imperial Hotel 111
INDEPENDENCE (Pitkin County) 199
INDEPENDENCE (Teller County) 116-117, 125
Independence Mine (Pitkin County) 199
Independence Mine (Teller County) 113, 116, 119
Independence Pass 104, 106, 131, 199-201, 205
Indiana Gulch 187, 188
Indomile Mine 221
Ingram, J. B. 279
Inter Ocean Mine 59
Inter-Laken Hotel 104, 108
Iowa 33
Iowa Mine 230
IRIS 158-159, 264
IRON CITY 139
Iron City Reservoir 139
Iron Mask Mine 213
IRONTON 243
IRWIN 149, 258-260
Irwin, Dick 149, 259
Irwin, George S. 255-256
Italian Mountains 265

J

J. J. Conway and Company Mint 179, 193
Jack's Cabin 258
Jackson Bar 31
Jackson County 221-222
Jackson County Map 220
Jackson, George Andrew 11-12, 31, 59, 62
Jackson, Sheldon 78
Jackson, William Henry 207
Jackson's Camp 62
Jacktown 102
Jacque, J. W. (Captain) 189
James, Jesse 172
JAMESTOWN 54-55
Japan Mine 279, 285
JASPER 162
JEFFERSON 83-84, 91, 191
Jefferson Creek 84
Jefferson, Thomas (President) 84
Jenny Lind Gulch 22
Jessie Mill 181
Johnson, Al 262
Johnson, Fred 262
Johnson, Jack 112
Johnson, John P. 228
Jones, Charles 188
Jones, John P. (Senator) 290
Judd, Garwood H. 260
Julia City 157
Jumbo Mill 181
Jumbo Mine 181
Junction House 205

K

Kasson, W. M. 137
Keebler Pass 260
Kellogg, S. B. 129
Kenosha Pass 84, 87, 88
Keokuk Hydraulic Mining Company 281
Kerber City 156
Kerber Creek 156
KEYSTONE 185
Keystone Science School 185
Killarney Kate 172

King Cole Mine 84
King, George E. 97
Kingston Peak 44
Kinney, Willis 234
KOKOMO 189-190, 197
Kreutzer Peak 251
Kreutzer, William R. 251

L

LA PLATA 294
La Plata County 294
La Plata County Map 293
La Plata Miner 228
La Plata River 294
Lace House 19
Lake, Arthur (Pastor) 99
LAKE CITY 231, 271-274
Lake City Phonograph 271
Lake Como 84
Lake County 96-108
Lake County Map 94
Lake Creek 106, 205
Lake Emma 235
Lake San Cristobal 272
Lakewood 58
LAMARTINE 46
Lamartine Tunnel 44
Lamb, Gus 250
Lampshire, Mary 50
Langrishe, Jack S. 17-18, 100, 179
Lant, David 218-219
Larimer County 71
Larimer County Map 70
Last Chance Mine 85, 169, 171
LaTourette, Charley 251
Lavery, Tom 100
LaVeta Pass 167
LAWSON 42-43, 50
Lawson, Alex 42
Lawson, Kate Coburn 42
Layton, Robert 38
Lead King Mine 262, 263

LEADVILLE 12, 57, 78-80, 83
  96, 97-101, 103-108,137, 156,
  188, 203, 205, 211, 255, 259
Leadville Area Map 95
Leadville Chronicle 99, 105
Leahy, Jack 199, 200
Learned, Henry 186
Leavenworth Creek 38
LEAVICK 78, 85-86, 92, 181
Leavick, Felix 85, 181
Lee, Abe 96-97
Lee, George S. 273-274
Leeper, John E. 34
Left Hand Canyon 55, 64
Left Hand Creek 53, 55
"Legend of Snow Blind Gulch" 255
LENADO 204
LIBERTY 156-157
Liberty Bell Mine 279-280, 284
Life of an Ordinary Woman, The 156
LINCOLN (Lincoln City) 180-181
Lincoln, Abraham (President) 80,
  177
Lincoln Creek 205
Lincoln Gulch 205
Lindstrom, Paul 35
Lion Creek 36
Little Annie Mine 161
Little Chicago 102
Little Ellen Mine 103
Little Giant Mine 228
Little Ida Mine 161
Little Jonny Mine 98, 102
Little Miami Mine 59
Little Pittsburg Mine (Little Pittsburgh)
  98, 105
Little Tycoon Mine 254
Logan Mine 60
LONDON JUNCTION 82-83
London Mountain 82
London, South Park and Leadville
  Railroad 82
Londoner, Wolfe 107
Lone Pine Mine 201

Los Pinos Indian Agency 272
Lost Canyon 129
Lost Canyon Placers 129
Lost Lake Camp 63
Lost Trail Creek 276
Loveland Pass 185
Low Line 112, 114
Lower San Miguel Mining District 281
Lucky 157
Lucky Strike Mine 264
Luis Maria Baca Grant No. 4 156, 158
LULU 224
Lump Gulch 24
Lytle, George 56
Lytton, Harry 181, 194

M

Machebeuf, Joseph P. (Father or Bishop) 18, 99, 136
Mackay, John W. 44
Mackey, Dick 23
Mackey Mine 23
Madonna Mine 132
Magna Charta Tunnel 257
MAGNOLIA 62
Magnolia Mine 62
Mahonville 137
Major Mine 161
MALTA 104
Mammoth Chimney Mine 264
Mammoth Mine 164
Mammoth Mountain 164
MANHATTAN 71
MARBLE 262-263
Marble City Times 263
Marble Mountain 151
Marshall Basin 288
Marshall Pass 137-138
Marshall, William (General) 138
Martin, William 56
Mary McKinney Mine 120, 126
Mary Murphy Mine 134-136

Masonic Lodges 20, 31, 39, 179, 189, 250
MASONTOWN 186
Masontown Mining and Milling Company 186
Masterson, Bat 171
Mastodan Mine 231
Matchless Mine 98
Mather, Charles 205
May Company 98, 111
May, David 98
May-Mazeppa Mine 255
MAYSVILLE 132, 138-139
Maysville Chronicle 138
McClancy Hotel 56, 65
McClellan Opera House 39
McCombe, Jack 99
McGowan, John (Dr.) 250-251
McIntyre, Charles 231
McKinley, William (President) 64
McLean, Edward B. 245
McLean, Evalyn Walsh 234, 244-245
Mead, Albert 273
Mears, Otto 234, 240, 242, 247, 282, 291
Meeker Massacre 12, 210
Melrose Gold Mining Co. 150
Mendota Mine 279
Meyer, Gus 24
Michigan Creek 84
Middle Boulder Creek 58
Middle Cottonwood Creek 131
Middle Park 21
Middle Park and Grand River Mining and Land Improvement Company 224
Middle Swan 191
Middle Swan River 191
Midland Terminal Railway 112, 114-117
Midway 112, 119
Mill City 32-33
Miller, Nick 73

Miller, Tom 179
Milligan, Tom 229
Miner Boy Mine 103
Mineral County 169-174
Mineral County Map 168
Mineral Hill 158, 264
MINERAL POINT 231, 262
Miners' Protective Association 73
Miner's Record, The (Denver) 82
Mining Register (Lake City) 271
Minnesota Mines Company 35-37
Minnie Mine 96
Miser's Dream 55
Mississippi River 11-12
Missouri Camp 211
Missouri Creek 22
Moffat 158
Moffat, David H. 57, 171
Mollie Gibson Mine 201, 202
MONARCH 132, 138, 142
Monarch Mine 132
Monarch Mining District 133, 138
Monarch Park 132
Monarch Pass 133, 138
Montana Theater 16-17
MONTEZUMA 87, 182-183, 185, 190, 195
Montezuma Mine 200
Montezuma Silver Mining Company 87
MONTGOMERY 79-80, 90
Morley 135
Morning Star Mine 255
Morris, Harry R. 251
Morrison, Robert M. 205
Morrison, Robert S. 36
Morrissey, John 99
Morse, Amos 36
Mosquito 80-81
Mosquito Creek 80, 82
Mosquito Gulch 80
Mosquito Mining District 81
Mosquito Pass 80-81, 83
Moss, John 294

Mount Aetna 131
Mount Carbon 265
Mount Champion 106
Mount Champion Mine 106
Mount Elbert 103
Mount Lincoln 80
Mount McClellan 37
Mount of the Holy Cross 212
Mount Princeton 130
Mount Princeton Hot Springs 130
Mount Shavano 133-134
Mount Sheridan 92
Mount Silverheels 79
Mount Sneffels 241
Mount Wilcox 37
Mount Wilson Placer Mining Company 281
Mountain Boy Gulch 106
Mountain Chief Mine 131
MOUNTAIN CITY 16-17, 25, 38
Mountain Daisy Cricket Club 20
Mudsill 85
Mudsill Mine 85
Mullen, Charles J. 96
Mullen, Joel K. 273
Murphy's Switch 135, 143
Myers, Julius 111
Myron Stratton Home 119

N

Nancy Gold Mine and Tunnel Company 61
National Belle Mine 241-242
National City 218
National Hotel 111
Neal, Mrs. Joseph 262
NEDERLAND 58, 67
Negus, Timothy G. 36
Neiman, Charles 218
Nelson, Olaf 234
Nevada Mine 282
NEVADAVILLE 20, 27
New York Cabins 214
New York Mountain 214

Newhouse, Samuel 99
Ni-Wot Mine 55, 56
Nickolls, George L. 34
Nilson, Sven 282
NINETY FOUR 45
Ninety Four Mine 45
Nobel, Alfred Bernhard (Dr.) 36
Nolan Creek 213
Northcentral Region, Colorado 14
NORTH CREEDE 170-171
NORTH EMPIRE 36-37, 48
North London Mill 81-82
North London Mine 81-82
NORTH STAR 255-257
North Star Mine 256
Northwest Region, Colorado 175
Nugget 24
Nunn, L. L. 283-284

O

O'Brien, Tom 262
Occidental Hotel 272
Ocean Wave Mine 274
O'Connell Tunnel 43
Odd Fellows 20-21
Ogden, Clara 271
Ogsbury, D. C. (Clate) 230
Oh My God Road 32
OHIO CITY 254-255
Ohio Pass 260
O'Kelley, Ed 171
Old Homestead 111
Old Lady Jack 261
Old Lout Mine 231
Old National Theatre 16
Old Ophir 282
Old Spanish Trail 281
Olney, Henry 251, 258
Omohundro, John B. (Texas Jack) 99
Only Chance Mine 264
OPHIR 282-283, 287
Ophir Loop 282
Ophir Pass 282

Ophir Station 282
Orchard, Harry 117
Oregon Creek 129
Orleans Club 171
ORO CITY 79, 96-97, 103
Orphan Boy Mine 80
Osgood, John Cleveland 206, 208, 263
O'Shanter, Tam 200
OURAY 105, 239-241, 243-246, 273, 284
Ouray County 239-248
Ouray County Map 238
Ouray Stage and Bus Co. 243
Ouray Times 239

P

Pacific Hotel 84
Pacific House Hotel 136
Packer, Alferd 272
Paepcke, Walter 203
Palace Manor 138
PANDORA 284
Pandora Mine 284
PARK CITY 80, 82
Park County 77-93
Park County Map 76
PARKVILLE 177, 179, 193
Parlin 254
PARROTT CITY 294
Parrott, Tiburcio 294
Parsons, John 81
Patrick, Daniel 190
Patsey Mine 36
Patterson, D. C. 59
Paul, J. Marshall 96
Pavich, Stella 272
Paymaster Mine 43
Payne's Bar 31
Payrock Mine 38, 41-42
Peabody, James (Governor) 115, 117, 280
Peabody's Seltzer 54
PEARL 222

Pearl Pass 259
Peck Gold Mining Company 36
Peck House 35
Peck, James 35-36
Peck, Mary Grace 35
Peekaboo Gulch 205
Pelican Mine 34, 38, 41
Pelican-Dives 38-39, 41
Pemberton 73
Pennsylvania Gulch 62
Pennsylvania Mine 184
Penrose, Spencer 112
Perault, Art Napoleon "Frenchy" 252
Peregrine, J. D. 57
PERIGO 21-22, 24, 28
Perigo Mine 21
Perry Mine 162
Peru 184
Peru Creek 184
Peru Creek Mining District 87
Peruvian Mine 184
Pharmacist Mine 119
Philadelphia Cattle Company 281
Phillips Mine 79
Phineas T. Barnum Circus 187
Phoenix Mining District 23
Pike, Zebulon Montgomery 11, 81
Pike's Peak 11–12, 111
Pilot, The (Crested Butte) 268
Pine Creek Mining District 23
Pinnacle Park 118, 125
Pioneer Mill 37
Pioneer Mine 36
PITKIN 252-254, 267
Pitkin County 199-208
Pitkin County Map 198
Pitkin, Frederick W. (Governor) 253
Pitkin Independent 253
Pitkin Mining News 253
Placer Creek 167
PLACERVILLE 281
Plaindealer (Ouray) 239
PLATORO 164
Pocahontas Mine 149

Poker Alice 172
Polar Star Mine 214
Poncha Pass 137
Poncha Springs 137, 138
Poor Woman Mine 61
Porphyry Mountain 54, 204
Portland Gold Mining Company 115
Portland Mine 113, 245
Poughkeepsie 234
Poverty Bar 218
Poverty Gulch 111-112
Powell House 152
PRESTON 181
Printer Boy Mine 96-97
Prunes (the burro) 78
Pry, Polly 272
Purcell, James 81

Q

Quartz Creek 254
QUERIDA 150, 153
Quigleyville 64

R

Racine Boy Mine 151
Ragland, Neil 225
Rainbow Lake 186
Rainbow Route 232, 243, 247
Rand 221
Raven Hill 126
Raymond Mine 255
Recen 189-190, 197
Recen, Andrew 190
Recen, Daniel 190
Recen, Henry 186, 190
Red Cloud Mine 54, 231
Red Elephant 43
Red Men 20
RED MOUNTAIN 241-244
Red Mountain City 242
Red Mountain Mining District 242-244
Red Mountain Pass 242-243

Red Mountain Town 242, 247
REDCLIFF 210-211, 215
REDSTONE 206, 208, 263
Redstone Inn 206, 208
Reed, George 182
Reese, Dempsey 228
Renniger, Theodore 171
Revenue Tunnel 241
Revenue-Virginius 241
Revett, Ben 180
REXFORD 190-191
Rexford Mining Corporation 190
Reynolds, A. E. 264
Reynolds, Jim 84
RICO 290-192
Rico Democrat, The 290
Rico State Bank 291
Rico Weekly Sun, The 291
Rink Opera House 203
Rio Grande County 161-162
Rio Grande County Map 160
Rische, August 98
Ritchie's Patch 129
Rivers, Harry 251
Roaring Fork Valley 202
Robber's Hill 172
ROBINSON 188-189, 196
Robinson Consolidated Mining Company 189
Robinson, George B. 188-189
Robinson, John 244
Robinson's Camp 188
Roby, H. W. 212
Rochester King 190
Rochester Lode 190
Rochester Queen 190
Rock Creek 261
ROCKDALE 140
Rocky Mountain National Park 224
Rocky Mountain News (Denver) 22, 78, 203
Rocky Mountain Sun (Aspen) 203
Rocky Mountains 16
Rogers, Buck 213-214

Rohwer, George 285
Rollins Gold Mining Company 22
Rollins, John Quincy Adams 21-22
Rollins Pass 21, 225
Rollins Steam Quartz Mill 22
ROLLINSVILLE 21-22, 24
Rombauer, R. E. (Judge) 36
ROMLEY 134-136, 143
Roosevelt, Franklin (President) 133
Roosevelt, Theodore (President) 112, 260
Rose's Cabin 274
ROSITA 149, 151, 153
Rough and Ready Hook and Ladder Team 18
Routt County 218-219
Routt County Map 217
Routt, John L. (Governor) 60, 61, 99
Royal Flush Mine 219
Royal Tiger Mines Corporation 182
RUBY 205-206
Ruby City 259
Ruby Mine 106, 205-206
RUSSELL 167
RUSSELL GULCH 20-21, 28
Russell Mining District 20
Russell, W. H. 36
Russell, William Green 11, 20
Rustic Camp 71
Ryan House 103
Ryan, Pug 178, 190

S

Saguache 272
Saguache County 156-159
Saguache County Map 155
SAINT ELMO 136-137, 254
SAINT KEVINS 105
SAINTS JOHN 87, 182-183, 185, 195
Salida 137, 141
SALINA 67
Sampson Mine 234
San Juan Chief Mill 231

San Juan County 228-237
San Juan County Map 227
San Juan Expositor (Eureka) 233
San Juan Herald 230
San Juan Mountains 129, 228-229, 272-273
San Juan Sentinel 239
San Luis Valley Land and Mining Company 157-158
San Miguel City 279
San Miguel County 279-288
San Miguel County Map 278
San Miguel River 279, 281, 283-284
Sanger Mine 162
Sangre de Cristo Creek 167
Sangre de Cristo Mountains 152, 157, 167
Santa Fe, N.M. 290
Savage Fork 285
Sawatch Range 257
Saxon Mountain 43
Scandia Mine 61
Scenic Line Band 132
SCHOFIELD 260-262
Schofield, B. F. 261
Schofield Pass 261-263
Scott, J. D. (Captain) 53
Sedgwick 156
SHAVANO 133-134, 138
Sheddon, Jack 188
Sheep Mountain 262
Sheep Mountain Tunnel 262
Shenandoah Mine 230
Sheridan Junction (Chattanooga) 242
Sheridan Vein 279
SHERMAN 271, 274-275, 277
Sherman Silver Purchase Act 12, 40, 101, 132, 152, 189, 231, 240, 275
Sherwood, Rupert 78
Short Creek 157
Short Line 112-114

Silver Bell Mine 282
Silver City 44
SILVER CLIFF 149, 151-152, 154
SILVER CREEK 43
Silver Islet Mine 253-254
Silver King Mine 183
Silver Lake 45
Silver Lake Mine 230
Silver Mountain 36
SILVER PLUME 37, 39-41, 50
Silver Plume Colorado 42
Silver Plume Jack Rabbit 42
Silver Plume Silver Standard 42
Silver World (Lake City) 271
Silverheels 79
Silverthorne, Marshall (Judge) 177-178
SILVERTON 228-231, 232-234, 236, 242-244, 273-274
Silverton Democrat 228
Silverton, Gladstone and Northerly Railroad 232-234
Silverton Standard 73
Simms, Cicero 77-78
Sisters of Charity 17
Six Mile House 42
Slate Mountain 213-214
Slumgullion Pass 272
Smith, Corydon Swan 182
Smith, George L. 171
Smith, Jefferson Randolph (Soapy) 99, 171
Smith, Joe W. 200
Smith, Sylvia 263
Smithsonian Institution 179
Smuggler Mine (Aspen) 201
Smuggler Mine (Robinson) 189
Smuggler Mine (Telluride) 279, 284-285
Smuggler Mountain 201
Smuggler-Union Mill 284
Smuggler-Union Mine 279-280
Snake River Mining District 87
Snare Creek 275

SNEFFELS 241, 246
Snow Blind Gulch 255
Snowden, Francis M. (Col.) 228
Snyder, Jacob 34, 41
Solid Muldoon (Ouray) 239, 242
Soule, Henry 234
Soup Bone Musical Club 256
South Arkansas Miner 139
South Arkansas River 133
South Central Region, Colorado 127
South Cottonwood Creek 131
South Evans Gulch 103
South London Mine 82
South Mountain 161-162
South Park 78, 84, 87-88, 137
South Park City Museum 78, 81, 86
South Platte River 11–12, 77, 80, 87
Southern Ute Indians 294
Southwest Region, Colorado 226
Spalding, Ruben J. 177
Spanish 157
Spanish Bar 31, 33
SPAR CITY 172
Spar City Spark, The 172
Spar Mine 201-202
Sparkhill 199
Spencer, George E. (General) 177
Spencer Mountain 62
Spottswood and McClellan Stage Line 187
Spurgeon, Elizabeth "Cockeyed Liz" 138
Squaw Gulch 120
St. Aloysius Academy 17
St. Cloud Mine 182
St. Elmo 134-135, 139
St. Elmo Mountaineer 136
St. Jacob's Mine 276
St. Louis and San Miguel Company 281
St. Louis Tunnel 103
St. Mary's Glacier 44-45
St. Patrick Mine 264

Staley, A. J. 239
Standard Metals 234
Stansell, J. B. 79
Star Gold Mining Company 36
Star Mountain 106
Stark, Annabelle 137
Stark Brothers' Store 136
Stark, Roy 137
Stark, Tony 136
Steele, Robert 38
Stein, Orth 99, 105
Stepler, Joseph 54
Sterling 275
Stevens, Anson P. 43
Stevens, Wallace 22
Stevens, William 97
Stockton, Ike 230
Stockton-Eskridge Gang 230
STONEWALL 135-136
Stonewall Mine 136
Stoney Pass 232
STRATTON 119
Stratton, Winfield Scott 111, 113, 116, 119
Stray Horse Gulch 102
STRINGTOWN 97, 102
Strong Mine 113
Stumpf, Joseph 103
STUMPTOWN 97, 103
Sturtevant, J. B. 64, 68
Sugarloaf 60
Sullivan, Jim (Deputy Sheriff) 230
Sullivan, John L. 100
SUMMERVILLE 63
Summit County 177-197
Summit County Map 176
SUMMITVILLE 161-162, 164
Summitville Nugget, The 161
Sunnyside 171
Sunnyside Mill 232, 236
Sunnyside Mine 230, 232-233, 235
SUNSET 60-62, 68
SUNSHINE 59-60
Sunshine Canyon 53

Sunshine Courier 60
Sunshine Mine 59
Sunshine Peak 274
Swan River 180, 182
SWANDYKE 191
Swandyke Gold Mining Company 191
Swickhimer, David 290
Swickhimer, Laura 290
Swilltown 104
Switzerland Trail Line (Colorado and Northwestern Railway) 61-62

T

Tabasco Mill 275
Tabasco Mine 275
Tabor 105
Tabor, Augusta 79, 97-98
TABOR CITY 105-106
Tabor, Elizabeth McCourt "Baby Doe" 19, 98, 200
Tabor Grand Hotel 100
Tabor, Horace Austin Warner 32, 57, 79, 96-98, 100, 129-130, 200
Tabor Opera House 100, 108
TARRYALL 77, 81-82
Tarryall Creek 81-82
Tayler, Charles 31
Tayler Waterwheel 31
Taylor City 105
Taylor, Frank M. 36
Taylor, Jim 250
Taylor Park 253
Taylor Pass 200-201
TELLER CITY 221
Teller County 111-126
Teller County Map 109
Teller, Henry M. (Senator) 17, 34, 183, 221
Teller House 16-18, 57
TELLURIDE 279-282, 285-286, 288, 291
Tellurium 275
"Ten Years War" 180

Ten-Mile Mining District 189
Tennessee Pass 105
Tenth Legion Mine 36
Terrible Mine (Ilse) 150-151
Terrible Mine (Silver Plume) 41-42
Terrible Silver Coronet Band 42
Terry, John 232
Teton 157
Texas House 100
Thom, William B. 210
Thomas, Lowell 114
Thomas, Otis C. 285
Thompson, George S. "Gassy" 184
Thomson, W. S. 229
TIGER 182, 194
Tiger Lode 182
TIN CUP 136, 139, 250-253, 258, 265
Tin Cup Pass 139
Tin Cup Record 251
Tintown 98
Tolland 22
TOMBOY 279-280, 285
Tomboy Gold Mines Company 285, 288
Tomboy Mine 279, 285
TOMICHI 257
Tomichi Creek 255
Tomichi Herald 257
"Tom's Baby" 181, 194
Tourtelotte, Henry 201
Tourtelotte Park 201
Tracy, Harry 218-219
Trail Creek 44, 46
Treasury Peak 262
Trout Creek 73
Trout Creek Pass 137
True Fissure, The 134
Truman, Harry S. (President) 138
Tungsten Camp (Steven's Camp) 58
Tungsten Mountain 58
Turkey Creek 210
Turner, Peter 59
Turner, Susie Sunshine 59

Turquoise Lake 105
TURRET 141
Turret Copper Mining and Reduction Company 141
Twain, Mark (Samuel Clemens) 32, 135
TWIN LAKES 99, 103-104, 108, 201
Twin Lakes Hydraulic Gold Mining Syndicate 129
Tyndall Mountain 150
Tyner 221

U

U. S. Forest Service 9, 200
Uncompahgre Mountains 239
Uncompahgre River 239
Union Mine 279, 284
Union Mining District 33, 36
Union Pacific Railroad 42, 219
Union Pass 38
Upper Swandyke 191
Ute City 201
Ute Creek 31, 46
Ute Creek Canyon 46
Ute Indians 12, 134, 228-239, 271-273, 294
Ute-Ulay Mine 271, 273
Utica Mine 55
Uzzell, Tom A. (Reverend) 99

V

Varney, A. J. 204
VICKSBURG 140
VICTOR 12, 113-114, 123, 147
Victorian Architecture 11-12, 19, 31, 102, 178, 259, 279
Vindicator Mine 115, 117
Virginia City 250, 258
Virginius Mine 240-241
Vivandiere Mine 141
VULCAN 264
Vulcan Enterprise 264

Vulcan Hill 264
Vulcan Mine 264
Vulcan Mines and Smelter Company 264
Vulcan Times 264

W

Waggoner, Charles 280
Wait, Lew 260
Waite, Davis Hanson (Governor) 115, 203
Wakely, G. D. 90
WALDORF 37, 49
Waldorf Mining and Milling Company 37
Wall Street Gold Extraction Company 61
Wall Street Journal 161
WALLSTREET 61, 69
Walsh, Thomas F. 105, 230, 234, 244-245
Wano Mine 54
Wapiti 180
WARD 55-56, 65
Ward, Calvin M. 55
Ward, Hiram 229
Ward Miner 56
Warman, Cy 170
Warrior's Mark Mine 188
Washington D.C. 98, 179
Washington Gulch 259
Watson, Joseph 39-40
Waunita Hot Springs 266
Way, George 218
Weaver 171
Webber, Henry (Mayor) 202
WEBSTER 87-88, 93
Webster, Emerson 87
Webster Pass 87, 183
Webster, William 87
Webster, William Wilcox 77
Wellington Mine 181
West Aspen Mountain 201
WEST CREEK 73-74

West Willow Creek 171
WESTCLIFFE 151-152, 154
Western Federation of Miners Union
    115, 117, 119, 273, 280, 284
Wet Mountain Valley 149, 150, 154
Wharton, J. E. (Dr.) 39
Wheel of Fortune Mine 241
Wheeler, B. Clark 201, 203
Wheeler Grand Opera House 202-203
Wheeler, Jerome B. 201, 202
White House Mountain 263
WHITE PINE 255-257, 268
White Pine Cone 255-256
WHITECROSS 271, 275
Whitecross Mountain 275
WHITEHORN 147
Whitehorn, A. L. 147
Whitehorn News 147
Whitlock, Logan 239
WIDE AWAKE 22
Wightman Creek 161
Wightman, James 161
Wightman, William 161
Wilcox 158
Wilcox, Edward John 37, 49
Wilde, Oscar 17, 100
Wilkerson Pass 86
Wilkinson, Burt 230
Williams, Bill (Parson) 81
Willow Creek 250
Willy, D. W. 182
WINFIELD (Chaffee County)
    140-141
Winfield (Teller County) (See Stratton) 119
Witting Hotel 147
Wolfe, J. H. 116
Womack, Bob 111
Wood, Alvinus 97
Woods, Frank 113, 147
Woods, Harry 113, 147
Woods Investment Company
    114, 118, 147
Woodstock 258
World War I 266
World War II
    35, 37, 81, 124, 179, 203, 232
Wright, A. E. (Dr.) 134

Y

YANKEE 45, 51
Yankee Blade Lode 130
Yankee Boy Mine 241
Yankee Girl Mine 242-244
Yankee Hill 32, 45
Yates House 221
Yellow Mountain 282
Yellow Pine Mine 60
Yule Creek 263

Z

Zweck, George 54

WORKS BY DAVE SOUTHWORTH

BOOKS: NON-FICTION

Famous Gunfights of the American West
Feuds on the Western Frontier
Colorado Gold Dust: Short Stories and Profiles
Colorado Mining Camps
Ghost Towns and Mining Camps of the San Juans
Gunfighters of the Old West
Gunfighters of the Old West II
Famous Gunfights of Texas
Leadville

BOOKS: FICTION

Franklin Hall
Rhymes of a Storyteller

VIDEOS

Colorado Mining Camps: A Pictorial Treasure of the Gold and Silver Boom
Leadville: The Boom Years
Mining Camps of the San Juans
Cripple Creek and the Mining Camps of Teller County
The Mining Camps of Northwest Colorado
Boulder County Mining Camps: A Look Back
The Mining Camps of Gilpin and Clear Creek Counties
The Mining Camps of South Central Colorado

AUDIO BOOKS

Gunfighters of the Old West
Colorado Gold Dust: Short Stories and Profiles
Billy the Kid and the Lincoln County War
Jesse James and the James-Younger Gang
Doc Holliday and the Earp Brothers

www.ingramcontent.com/pod-product-compliance
Lightning Source LLC
Chambersburg PA
CBHW071218080526
44587CB00013BA/1416